What Works for GE May Not Work for You

Using Human Systems Dynamics
to Build a Culture of Process Improvement

What Works for GE May Not Work for You

Using Human Systems Dynamics to Build a Culture of Process Improvement

Lawrence Solow and Brenda Fake

CRC Press
Taylor & Francis Group
Boca Raton   London   New York

CRC Press is an imprint of the
Taylor & Francis Group, an **informa** business

A PRODUCTIVITY PRESS BOOK

Productivity Press
Taylor & Francis Group
270 Madison Avenue
New York, NY 10016

© 2010 by Taylor and Francis Group, LLC
Productivity Press is an imprint of Taylor & Francis Group, an Informa business

No claim to original U.S. Government works

Printed in the United States of America on acid-free paper
10 9 8 7 6 5 4 3 2 1

International Standard Book Number: 978-1-4398-2599-0 (Hardback)

Library of Congress Cataloging-in-Publication Data

Solow, Lawrence (Lawrence A.)
    What works for GE may not work for you : using human systems dynamics to build a culture of process improvement / Lawrence Solow and Brenda Fake.
        p. cm.
    Includes bibliographical references and index.
    ISBN 978-1-4398-2599-0
    1. Organizational effectiveness. 2. Organizational change. 3. Process control. 4. Quality control. I. Fake, Brenda. II. Title.

    HD58.9.S66 2010
    658.4'013--dc22   2010000672

Visit the Taylor & Francis Web site at
http://www.taylorandfrancis.com

and the Productivity Press Web site at
http://www.productivitypress.com

This book is dedicated to my parents, Paul and Sheila Solow.
They provided the DNA, motivation, and solid foundation
that made my contribution to this book possible.

Larry Solow

This book is dedicated to my clients across industries and
companies for enriching my work and understanding of the
world. Thank you for the experiences and validation that
people are the most important asset of any organization.

Brenda Fake

Contents

SECTION II   So What?

SECTION III   So What?—Take 2

SECTION IV Now What?

Foreword

As founder of the field of human systems dynamics (HSD), my mission is to build capacity for individuals and organizations to be effective and productive, even when they cannot predict or control the future. Applying complexity, chaos, and other nonlinear sciences to human systems—individuals, teams, and organizations—HSD provides a new way to address critical needs in today's challenging global environment.

Nowhere is the opportunity for this approach greater than in the area of process improvement. People everywhere strive to do better, be better, and achieve better results for the organizations to which they commit themselves. There is nothing more heartbreaking than to see these efforts either fail completely or only partially achieve their potential.

Larry Solow and Brenda Fake bring the best of three different worlds to this book. First, both have invested their time and energy in understanding the concepts, models, and tools of HSD. Second, both have first-hand knowledge of process improvement. Larry is a Six Sigma Black Belt and has created his own innovative problem-solving process. Brenda has worked with world-class organizations to understand the human side of process improvement and organizational change. Finally, both are seasoned, experienced change agents. They have worked with a wide variety of organizations that span profit and nonprofit, industry segments, and geographies. They write from first-hand experience, sharing their stories of what has and has not been effective in making change happen.

Throughout their book, Larry and Brenda emphasize that adaptive action is not intended to replace existing process improvement tools, but rather to complement them. I could not agree more. Viewing organizations as complex adaptive systems invites focus on patterns and the recognition that influencing critical dimensions of these patterns creates a shift in them. It is a different mindset than the mechanistic, cause-and-effect thinking that permeates so much of today's leadership thinking. Not better and not worse; there are many places where viewing systems as linear is both true and useful. It is in those other places, where flexibility, rapid change, and the need for adaptive action are called for, that HSD-inspired constructs and tools shine.

I found *What Works for GE May Not Work for You: Using Human Systems Dynamics to Build a Culture of Process Improvement* to be theoretically sound, practical, and easy to read and understand. Blending traditional models and new thinking, the authors have certainly provided food for thought for those struggling to make process improvement a way of life in their teams and organizations. I believe these ideas and tools can help you build capacity for yourself, your team, and your organization to respond creatively and effectively to complex change.

Glenda H. Eoyang
Founding Executive Director
Human Systems Dynamics Institute
Circle Pines, Minnesota

Preface

If you can imagine a system ... comprised of individual agents ... each with the freedom to act in unpredictable ways ... yet whose actions are interconnected, but not known to any of the agents in the system ... then we can begin to explain how this book evolved. From a strictly linear, "cause and effect" standpoint, the chances of Brenda Fake and Larry Solow joining forces was astronomically small, yet it happened. Three people—three seemingly unrelated stories—came together and resulted in the story you are about to read.

We begin with Dr. Glenda Eoyang's story. Her groundbreaking work in applying nonlinear dynamics to human systems created the field of Human Systems Dynamics (HSD). It was HSD that would ultimately serve as the catalyst for Larry's and Brenda's collaboration on this book.

DR. GLENDA EOYANG

Eoyang never outgrew the two-year-old's question, "Why ...?" Growing up in a string of small towns in West Texas, she wondered why a church continued to exist even after the building burned down. She wondered why one little community celebrated different holidays than others. She wondered why sports were a passion for some, a pastime for others, and an irrelevancy to her. She wondered why some people could build words and sentences into a work of art and others barely made themselves understood. She wondered.

St. John's College in Santa Fe, New Mexico, turned her curiosity toward classics of the Western World. In reading and discussing great works, she discovered that Euclid, Plato, Aristotle, Galileo, and Newton also wondered, and that they did it very, very well. Studying history and the philosophy of science inspired her to reflect in new and surprising ways on her own experience with individuals and groups. She taught physics and chemistry, designed early computer-based training, started her own companies, and supported others as they implemented computer systems and the management systems that supported them. Her intuitions about learning and groups worked most of the time, but when they didn't, she wondered why.

In 1986 she happened upon James Gleick's work, *Chaos: Making a New Science,** which was one of the earliest explorations of chaos and complexity science that was accessible to a lay audience. Though the book was about nonlinearity in physical systems, she saw it as a source of insights about the social systems around her. In the decades that followed, Eoyang learned everything she could about nonlinear dynamics and practiced using those insights to support herself and others as they engaged with complex challenges. In 2003, she founded the Human Systems Dynamics Institute to "develop theory and practice at the intersection of social and complexity sciences." Today she teaches internationally and leads a network of 135 scholar–practitioners who continue to explore the dynamical interactions of people and the physical systems that influence and are influenced by them.

We leave Eoyang's story here, as she continues her mission to deepen the knowledge and application of HSD. You will see how her story intersects with Larry's and Brenda's.

LARRY SOLOW

Larry has always been a person to ask questions and challenge traditional theory. His M.O. for learning is to be introduced to a new subject, discuss it, and, during the conversation, identify assumptions, inconsistencies, and exceptions.

As a freshman at Rutgers University in 1974, Larry enrolled in a course named "Introduction to Human Communication" taught by David ("Louie") Bender. Bender's common

---

* Gleick, James. *Chaos: Making a New Science*. 1987. Penguin Books: Middlesex, U.K.

sense, provocative approach to learning captivated Larry, and he chose to major in human communications. He went on to get his master's degree in organizational communication and eventually found a job with Robert Bell and Company, a short-interval scheduling consulting firm. Reflecting on that experience, Larry laughs as he recounted almost being fired several times because of his incessant questioning.

After a stint with another consulting firm, Larry joined Harley-Davidson, Inc. in York, Pennsylvania, as a senior OD consultant. Harley provided a powerful "learning laboratory" for his continuing questions about the intersection of theory and practice. He was exposed to the concepts and tools of Total Quality Management (TQM) during this time.

A consulting colleague, Lisa Marshall, knew of Larry's questioning nature and suggested he contact Mark Michaels, chair of the Chaos Network. Michaels suggested Larry read the same *Chaos: Making a New Science* text that was so instrumental in Eoyang's thinking. That conversation led to an invitation to present at the first annual Chaos Network Conference in Washington, D.C. That is where Larry met Eoyang for the first time.

Their paths diverged as he moved to a division of AlliedSignal Aerospace as a manager of Total Quality. Here he would continue to refine his thinking regarding the "messy" intersection of process improvement tools, change, and organizational culture.

Larry left AlliedSignal in 1993 to start his own consulting practice. He earned his Six Sigma Black Belt in 2004. Working as a subcontractor for various local colleges, Larry helped implement Six Sigma processes in 10 small- to mid-sized companies over the next several years, with varying degrees of success. Never having lost his questioning nature, he asked, "If the same instructor is teaching the same way, using the same materials, in the same sized companies, how come the results are so different?"

His search for the answer led him back to chaos and complexity theory—and reconnecting with Glenda. She invited him to participate in the HSD professional certification process, and he was certified as an HSD Professional in 2008. Conversation and more questions led Larry to ask her who else might have a similar interest in the intersection of process improvement and HSD. One of the names she shared was Brenda Fake.

BRENDA FAKE

In Brenda's case, the burning need to ask "why" questions to everything existed early in life—so much so that she was told to stop asking so many questions. Later in life, she was introduced to a tool called the "5 Whys" used to surface the real issues on a team or project. Brenda laughs as she recounts "feeling somewhat vindicated" for her persistent questioning.

However, a stronger, more powerful force in Brenda's world view was seeing with a great deal of clarity that all things are more connected than not: religion, science, people, politics, paradox, irony. Why didn't others understand this connectedness? While at the University of Minnesota as a communications major, she secretly hoped some class would confirm to her that universal laws exist and could explain how the world worked. She reflects, "I was not as concerned about the differences and the details, but rather the patterns of similarities. It seemed if everyone could see the similarities, all of the ills of the world, work, and relationships could be solved."

Around the time she left college to enter the working world, complexity theory was just in the early stages at the Santa Fe Institute, though she did not learn about it for another 10 years. Shortly after receiving her MBA, a colleague recommended the book, *Complexity: The Emerging Science at the Edge of Order and Chaos** by M. Mitchell Waldrop. Brenda said, "I like books, but for me to actually read one cover to cover and devour every page is unusual. I loved the story of the Santa Fe Institute's summer school, with the greatest minds from a variety of hard and soft sciences across several fields of study coming together to not talk differences, but similarities and connections. Vindication. I am not alone in thinking stuff is connected!"

---

* Waldrop, M. Mitchell. *Complexity: The Emerging Science at the Edge of Order and Chaos.* 1992. Simon and Schuster: New York.

As Brenda's career progressed and life experiences unfolded, she continued to see connections and worked to help others do the same. In her work of improving business processes, especially as it related to managing the impact of change, this perspective provided real value to her clients.

Brenda had become a consultant in Organization Effectiveness working with engineers in public works and manufacturing, where she met Glenda Eoyang in 2002 on a strategic planning research project for the public sector. The project frustratingly started with reviewing books and articles on strategic planning, which made sense, but did not reveal any new thinking. Each article or book pointed out a unique part of a planning process, but overall they all followed a rather common linear approach. All the rhetoric was very fixated on the plan, leaving managers or leaders stuck with a useless document. Brenda remembers thinking to herself, "So what if the shit hit the fan in the middle of plan execution? The 'plan' is useless! None of the reviewed materials seemed very adaptive. This was the problem we needed to solve for this project. Our task was to outline a process for application that would actually help Public Works leaders."

In her initial conversations with Eoyang, Brenda knew she had met someone who had a line of sight on how to reduce the noise of all the rhetoric, find the simple patterns and connections, adapt, and leverage them for the bigger picture. Brenda says, "I can't say what the exact exchange was in the meeting that told me to follow up with Glenda. It was the sense of knowing when you have met someone who will have a connection to your thinking and your work. I was sure she would understand my need to find interdisciplinary patterns and give words to my secret theories."

Further work on the project together led to Brenda's completing the HSD Professional certification early in the history of the institute in the summer of 2003. The result was a confirmation that the patterns of the big picture were, in fact, connected by the separate theories across time and space. People are, in fact, more connected than not, and the HSD work helps give meaning and definition to her fuzzy concepts.

Brenda moved on to Honeywell Aerospace, where she introduced the HSD work to a group of engineers to help them sustain their efforts in applying process improvement in new product introduction. Eoyang suggested that Brenda talk with Larry, whom she knew shared her interests and background in Six Sigma, in HSD, and in helping clients create significant change, so they can reap significant gains on their consulting investment.

Three different stories from three different people, each starting at different points, with different career trajectories, who weren't connected in any meaningful way. Yet their stories intersected and the result is this book—and the application of chaos and complexity theory as they apply to implementing process improvement in human systems.

Acknowledgments

There are many people who helped make this book possible. Together, authors Brenda Fake and Larry Solow would like to acknowledge the following people: First, we thank Dr. Glenda Eoyang. Her Human Systems Dynamics (HSD), provides one of the foundations of our work. In addition, Glenda graciously gave of her own valuable time and writing talent to help us fine tune our message and assure the accuracy of our HSD-related content. Any mistakes or omissions in that regard are our responsibility, not hers.

Julia Herzing and Royce Holliday, members of the HSD Institute, were gracious in their support of our effort. Royce, thank you for allowing us to share your 7 Cs model with the world.

A big thank you to Productivity Press and Michael Sinocchi for seeing the possibilities in the book; to Lara Zoble who provided guidance during the writing process; and to Jay Margolis for your collaborative approach as production editor, resulting in a well-done book and smooth production. It was our good fortune to have crossed paths with all of you.

We also express our heartfelt appreciation to our readers. Individually and collectively, John Vic Grice, Tina Kontos, Jennifer Koschmann, and Gregory Solow (listed alphabetically) provided extraordinarily valuable input from a variety of perspectives.

Lois Lenox, you are the best. We cannot thank you enough. Your assistance with the mechanical details made it possible for us to concentrate on the actual writing of the book.

In addition to the above, Larry would like to acknowledge the following people: Members of the ConCenter Alliance—Karen Bading, Janet Crawford, Charles Feltman, Dee Kinder, and Lisa Marshall. Individually and collectively, you have been partners in his personal and professional development for a long time. In many ways, it was your encouragement to "think big" that resulted in this book.

Larry would also like to acknowledge Warren Wilhelm, whose support over the years has been invaluable. Larry has been honored to collaborate with you on a variety of projects, many of which have formed the ideas Larry brought to this work.

Bob Kautz, as Larry's Six Sigma mentor, you put up with his incessant questions with patience and good humor. Your ability to distill complicated concepts into language that nonstatisticians can understand and use is a direct result of our conversations.

Larry would like to acknowledge the contributions of his past clients. While all have added to his base of experience, several organizations and leaders have played a special role in catalyzing his thinking about HSD and process improvement. Michael Ruggieri at Comar, Michael Eyler at Wheaton Science Products, John Nawrath at Tyco, and Rudolf Messinger from UNICEF—thanks to you all.

Leaving the best for last, Larry thanks his wife, Maureen, and sons, Gregory and Ryan. As big an accomplishment as this book is for him, it pales in comparison to the wonder of having all of you in his life.

Brenda would like to acknowledge: Vicki Poels who has played many important roles in her life—mentor, teacher, manager, colleague, golf partner, and, most importantly, her lifelong friend.

For their generosity of time, personal coaching, and professional insight into the unknown world of publishing, Brenda's deepest gratitude to Valerie Dow and Beth Wagner-Brust.

To the San Diego State University Writers Conference. It is the best investment in learning the business of publishing. Please keep up the great work of bringing fledgling authors with hope, together with experienced publishing professionals; the vibe is magical.

Sherry MacAloon, Brenda's fifth grade teacher, who encouraged the questions, focused the abundant energy, and introduced her to science.

Lastly, to her husband, Tom Rushfeldt, who encourages Brenda to think big and take risks in love, life, and work. He truly is the best.

Introduction

The journey to this book has been a challenging road of experience and observations. It is the result of witnessing clients honestly and earnestly attempting to implement process improvements; trying to make a positive difference for customers, employees, and shareholders.

We, the authors, with more than 50 combined years of change management and process improvement consulting experience, noticed four separate and distinct patterns in our client engagements. First, all of our clients implemented business process improvement using linear approaches. Using these methods makes sustainable improvement almost impossible to achieve. The lack of ability to adapt and get past the many steps and rules of Six Sigma or Lean implementation "thou shalt" strategies was evident in this pattern.

Second, we observed managers and leaders with advanced college degrees, lots of "book smarts," and enthusiasm who could not translate their knowledge into effective, sustainable actions that positively influenced change.

Third, we recognized that process improvement implementations didn't seem to last. After an initial burst of program-like energy and activity, organizations and processes reverted back to the same old way. The implementation effort was unable to sustain itself over time. The goal of changing the culture—the daily way of working within the organization—to make process improvement part of daily organizational life did not reach its full potential. This was also evident in the inability to connect improvement efforts back to cash flow and real impact to the bottom line.

The fourth pattern we identified was how often leaders failed to think outside the recipe for success defined by the most recent Six Sigma or leadership books. One of the real challenges of effective leadership is the translation and adaptation of theory to the real world environment in which leaders and managers work every day.

An example of this pattern occurred when Brenda Fake arrived at one of her client engagements to be greeted at the beginning of a meeting with the question: *"Have you read the latest book from so and so? It is the best one I've read. In fact we are all reading it because it tells us how we should really do business.*" The text was usually written by a successful former leader from General Electric (GE). Because it was difficult to argue with the success of GE (especially in the 1980s and 1990s), the author's concepts *must* be the way to business success. Those ideas were clearly helpful to GE at the time the book was written. The risk comes in assuming it can be "cut and pasted" into other companies without a great deal of modification.

To be clear, we are not writing this book to discuss GE—the naughty or the nice. We do intend to provide information and tools to help readers do two things: (1) help translate new, relevant theory to their own management practices, and (2) provide new resources for managing and sustaining process improvement in today's complex nonlinear environments.

This book was born from the authors' frustration and aggravation at seeing too many process improvement initiatives being implemented and promoted as corporate culture change. Based on the idea that any new business improvement effort will be the new competitive way of doing business, Six Sigma or Lean Enterprise is often identified as *the way* to lead the improvements and change the culture.

Now add rhetoric from the organization's leaders and managers stating that everyone will be involved in the decision process. Yet, in reality, many of the decisions about if and how to implement these new tools have already been predetermined and formulated. Typically, these approaches for implementation were taken from a book or consultant who did not highlight the many variables of the environment that enabled the "successful" effort to emerge.

Corporate politics also play a role. Employees feel they are forced to "drink the Kool-Aid"" of this version of process improvement or suffer negative consequences. After exposure to the "new" set of process improvement (PI) tools, they realize the new concepts are what good engineers, operations professionals, and others with a degree of common sense already do. *"This is not magic. Why are we making a big deal of it?"*

Another frustration arises from an organizational disconnect. Power plays are alive and well. Process improvement functions have become institutionalized in organizations and seek to justify their existence and expand their role. Without direct profit and loss accountability, PI groups are a support function; "overhead" on financial statements, to be reduced like any other burden cost in times of stress. As a result, their efforts often fail to link to organizational strategy and priorities as they protect their positions and influence.

We see many organizations build internal, stand-alone PI departments to deliver the skills and knowledge of Six Sigma, Lean, or Quality to an organization. On the one hand, creating a separate PI function means their efforts are less likely to be diluted. On the other hand, it is easy for these groups to be disconnected from the business, contributing to the overhead perception. There is also the risk of diluting the process improvement effort because PI resources are often reassigned to emergency problems. This results in longer term improvement efforts being set on the back burner of priorities.

Another issue is that these teams are comprised of members selected for their technical and analytical skills. Many times, they are not provided the basic consultation skills that help build customer-driven relationships and a supportive client base. The "technical" experts can fall into the "I have the knowledge" and "I must be right, after all, I'm the expert" traps contributing to the disconnect.

Another frustration is watching organizations attempt to keep a process improvement effort alive following a change in leadership. One leader starts something and leaves, followed by another leader who has a different set of priorities. It is no wonder whole organizations just hunker down and avoid these process improvement initiatives. Why do the hard work of change when it will be reversed by the next administration?

Finally, we are frustrated by the inability of organizations to recognize that change happens differently in human systems than in mechanical systems. The need to adapt (rather than strictly copy) an approach to be effective in the particular context of an organization is ignored and the deck is stacked for failure.

Having vented our frustrations, we move on to other business challenges.

The authors of this book are doers, practitioners in the field of organizational change and improving effectiveness. We've had our share of successes and developed some calluses and scar tissue from change efforts that didn't work so well.

We believe in the maxim: *"There is nothing as practical as good theory."** Useful theories provide ideas and expectations about what and how things work in the real world.

---

* Attributed to Kurt Lewin in Marrow, A.J. (1977). *The Practical Theorist: The Life and Work of Kurt Lewin*. New York: Teachers College Press. (Originally published by Basic Books, New York, 1969.)

Joel Barker, futurist and author, states, "Vision without action is merely a dream.* Action without vision just passes the time. Vision with action can change the world." Our variation is: "Theory without action is only interesting. Action without understanding is hoping to get lucky. Action based on theory provides the best chance for learning and success."

Another name for theory-based practice is *praxis*. Merriam-Webster Online defines praxis as "the practical application of a theory."† It is in praxis where leadership starts; it is how individuals make sense of their environment and how that information is translated into creative applications to make change happen.

Here are some specific ways we hope to blend theory and practice throughout this book:

Plain language: In our experience, people who are charged with getting things done don't have a lot of tolerance for "high falutin'," fancy language and jargon. They want the message straight, delivered in a language that can be easily understood and communicated to their people. We will do our best to minimize the three paragraph, graduate school syntax we sometimes find in other books. *"We think …"* is preferable to *"It could be argued that … "*.

Translation: The authors share the ability to translate complex theoretical and mathematical constructs into useful tools. When Larry teaches Six Sigma Green Belt training, he doesn't provide his participants with a calculator. Rather than potentially lose people in the mathematical calculation of a standard deviation, he shares the reasons for using one, the assumptions that underlie them, and how they inform understanding and decision making.

Stories: We are fortunate to have the real world as our laboratory. Our experience as employees, internal change agents, and external consultants has provided us with experience in well over 100 organizations. We will share our first-hand experiences—the ones that worked and those that didn't. We will not use specific client names (to protect the innocent—or guilty), but will provide an idea of industry and scale (one size does *not* fit all in our world).

---

* Joel Barker video, *The Power of Vision*. Star Thrower Distribution, Inc.
† Merriam-Webster Online Dictionary, http://www.merriam-webster.com/dictionary/praxis (accessed August 11, 2009).

Results: Results are where the "rubber meets the road" for practitioners. *"The theory is all well and good, but did it work?"* is an essential question that we will answer in our stories and examples.

Context: We are not big believers in "one size fits all." Stories, theories, successes, and failures need to be understood in context. That context includes the organization's industry, history, competitive environment, the technology available to them, political party in power at the time, etc. We'll try to be clear in the stories and anecdotes we share about these kinds of issues. To the extent that you, a fellow practitioner, say, *"But we operate differently,"* GREAT! Adapt our ideas to make them applicable, or simply recognize them as not useful and move on.

Articulating our underlying assumptions provides an additional function. It allows them to be questioned or challenged. If you don't agree with our stance, that's fine. Identify your own stance, assumptions, and biases. Share them with your co-workers. Let them know where you're coming from.

Having shared out writing approach, we move forward to the book itself. *What Works for GE May Not Work for You* is divided into four sections. The naming convention is based on a Human Systems Dynamics (HSD) concept called *adaptive action*, which will be explained later. The first section we refer to as *"What?"* Here, we provide descriptions of the key bodies of knowledge we utilize throughout the book. In the first chapter, we reaffirm the business case for a different look at process improvement. In service to transparency, we also explain the simple rules we followed during our writing process.

In the second and third chapters, we provide a brief history and summary of two recent process improvement processes, Lean Enterprise and Six Sigma. After providing this shared foundation with our readers, Chapter 4 introduces the HSD discipline, which is the body of knowledge that provides the foundation for much of our thinking and recommendations.

The second section is titled *"So What?"* Having presented our descriptive view of Lean, Six Sigma, and HSD, what are the issues? What are the opportunities? This section introduces the case story of *TryinHard Marine*, dramatizing and highlighting the dynamics of a typical process improvement implementation. After Chapter 5 introduces the section, Chapter 6 provides an introduction to the company. Chapter 7 highlights the dynamics of a typical project start-up, while Chapters 8 and 9 focus on the dynamics of the training element of process improvement implementations. Chapters 10 and 11 move forward in time to describe the future consequences of a potential linear approach to implementation.

In the third section, titled "*So What?—Take 2*" the story of *TryinHard Marine* is retold. In this second telling, efforts are made to deal with a range of complex, nonlinear, and emergent organizational issues as they arise during the implementation and evolution of the Six Sigma initiative. Chapter 12 introduces the section. Chapters 13, 14, and 15 highlight new conversations and tools available for choosing to move forward with a Six Sigma initiative, selecting an outside business partner, and project selection. Chapter 16 revisits the training component, providing the means for a direct comparison with the approach highlighted in Chapters 8 and 9. Chapter 17 provides a new perspective for dealing with the change dynamics that are an inevitable part of a major process improvement effort. Chapters 18 and 19 once again move the process forward in time in order to highlight the long-term implications of an adaptive approach.

The fourth section of book asks, "*Now What?*" Here we will more fully outline the tools that have been introduced in the case stories and describe where and when they might be applicable. We look more deeply at the areas of assessing readiness to begin a process improvement initiative, selecting consultants and internal "belt" candidates, and tools for project selection. Additionally, we introduce additional tools and concepts to enable "adaptive action" at all levels of the organization.

The final chapter of the book provides a summary of our thinking as well as additional resources for your continued learning.

A final thought. This book is not intended to be an indictment of Jack Welch, GE, or anyone who has written a book on management and leadership practices. Instead, by adding new concepts, models, and tools, we hope readers can use them to improve whatever business improvement process they are currently using. The key is awareness of your own practices, setting conditions for success, and learning to adapt them in a way that is effective in your world. We hope you will find the information we provide helpful in thinking and behaving in adaptive ways that can be applied in your work and organization.

Section I

What?

1

Introduction to "What?"

Between the authors, we've seen way too many process improvement (PI), continuous improvement (CI)*, and Lean Six Sigma (LSS) efforts either fail completely or achieve only a small fraction of their potential. Organizations invest a phenomenal amount of time and money on PI training and projects. These suboptimal results are painful from every perspective.

CONSEQUENCES OF FAILURE

Busy employees already invest their energy in the hard work of learning new skills, language, and behaviors. They suffer through the learning curve of putting their new knowledge and skills into practice, dealing with the inevitable setbacks and embracing the new paradigms espoused by their leaders and trainers. Who can blame them for being frustrated when their hard work and ideas are never put into place, or are implemented and then "fade away quietly into the night."

This is not the fault of the process improvement tools themselves. These issues are a result of failing to understand how to create conditions that will help sustain improvement efforts over time and provide the desired results.

Managers and leaders suffer as well. They have to deal with the "opportunity" cost of the time and money invested in the LSS effort. When these efforts falter or fail, leaders can only shake their heads about the more productive ways those scarce resources might have been used. They must also

---

* We will use the terms continuous improvement (CI) and process improvement (PI) loosely and interchangeably. Purists will argue that PI is more local and focused; that CI is more broad-based and ever-present. At least here, we don't feel a need to differentiate.

deal with the frustration, aggravation, and general negativity that failed improvement efforts generate for all stakeholders.

Negative past experiences and failed efforts leave behind a residue of bad taste and painful memories. Figure 1.1 illustrates the dynamic. Leaders enter into the process improvement effort with some level of credibility and trust. Their initial "cheerleading" results in a temporary increase in credibility as they passionately communicate that "this time it will be different." As the organization fails to successfully implement or sustain the new culture and tools, the failure is seen as yet more proof that leadership doesn't "walk its talk." As a result, both management's credibility and the process improvement effort are worse off than before the LSS implementation began. The lack of results creates a belief that change is hard. As this belief is enforced, it can become ingrained into the organization's culture.

Customers suffer in multiple ways. First of all, they suffer because responsiveness by the organization to their needs slows because employees are in training and meetings. As a result, they are not doing their regular work.

Secondly, customers suffer because improved processes would benefit them. When a project designed to reduce scrap fails, the customer can't take advantage of (the potential for) higher quality, better throughput, and lower prices. When a Lean project designed to reduce cycle time fails, the customer pays with slower response from their suppliers than was possible or necessary. Expectations were raised and not met, resulting in a loss of the customer's trust. This is a similar pattern to the loss of credibility suffered by the organization's leadership shown in Figure 1.1.

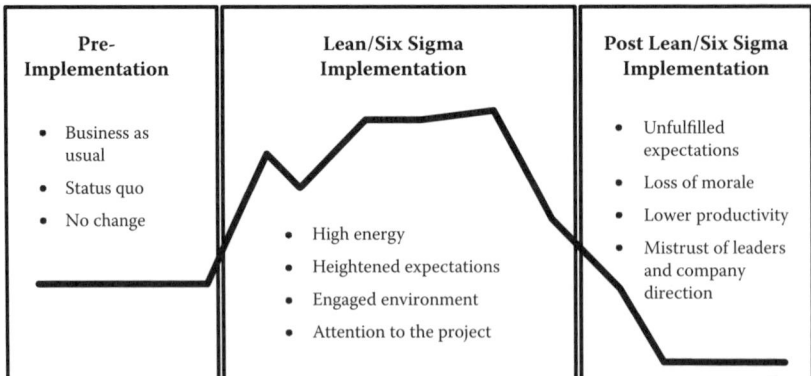

Pre-Implementation	Lean/Six Sigma Implementation	Post Lean/Six Sigma Implementation
• Business as usual • Status quo • No change	• High energy • Heightened expectations • Engaged environment • Attention to the project	• Unfulfilled expectations • Loss of morale • Lower productivity • Mistrust of leaders and company direction

FIGURE 1.1
Prepost implementation stages. (Courtesy of Brenda Fake.)

Thirdly, customers suffer because they deal with employees who are less engaged, less energized, and less resilient. Wouldn't you feel that way as well if you watched your hard work and investment fail? Think of going to a local fast food restaurant. Upon entering, the sounds of the manager yelling at the employees are clearly audible. Workers are shaking their heads in anger, disbelief, or resignation. They don't smile, look you in the eye, or correctly take your order. Will you return to that restaurant, especially since there are so many other choices available nearby?

Stockholders are another group impacted by the success or failure of process improvement efforts. Investing hard earned, discretionary dollars in the organization, they expect a reasonable return on investment. Lean Six Sigma implementations are often high dollar undertakings—investments that often seem to be flushed down the drain. As in the restaurant example above, reasonable investors have alternatives when it comes to where to invest their money. How likely would you be to invest in an organization that squanders your money?

As we mentioned earlier, and perhaps counter-intuitively, we *don't* think the problem is with the process improvement tools themselves. We believe that concepts and techniques like Six Sigma and Lean are strong and useful. So, if they aren't the problem, what is the source of the angst? *"If it ain't the what, it has to be the how,"* and the *"how long."* We believe that PI tools and methods are fundamentally sound; however, the way they are implemented and sustained is not.

DON'T IMPLEMENT LSS FOR THE WRONG REASONS

One set of issues is related to the fact that LSS processes are implemented for the wrong reasons. The failure modes listed below make it extraordinarily difficult for a new culture to take hold. Some of the flawed business motives for process improvement implementation include:

It's the right thing to do; it sounds good—Continuous improvement methods are often implemented because they sound like the right thing to do. Who wouldn't want to claim to be doing something with the word *improvement* in it? As a result, missions and visions are populated with phrases like: *"We will constantly strive to be better in everything we do."* So far, so good. The problems arise when the follow-through is haphazard.

This is a failure in the implementation of a process improvement plan or initiative. Several of our clients ask employees to sign a copy of the mission or vision statement signifying their commitment to making it a reality. Unfortunately, sometimes that's the only thing done.

We need to do something different—When desired results are not achieved, leaders often feel they need to do something differently. The expected end results are usually quantified and clear. When a new process improvement initiative doesn't create the desired results, it is often easier to replace it with a new one—any one—rather than try to identify the underlying root cause of the failure.

Others are doing it in the industry, so it must be good—Folk wisdom from the past advise: *"You'll never get fired if you hire IBM,"* even if IBM products and services weren't the best fit for the situation. Leaders hear a competitor initiated a Lean or Six Sigma program and the juices start flowing. *"We need to do that, too! If we don't 'keep up with Brand X,' we'll be left behind."* The word comes down that the organization will immediately implement an even better Lean Six Sigma program.

The title of this book was selected to draw focus to this particular concern. The companies where some of these methods began (Motorola, General Electric, AlliedSignal (now Honeywell), Boeing, 3M, Lockheed Martin, and other Fortune 500 organizations have been undeniably successful implementing change using LSS tools. The intent of this book is to offer those implementing LSS with information that will help them identify options for action to increase the odds for a successful and sustainable process improvement implementation in their own organization. With vast amounts of financial and human resources, GE and others made the investment to budget thousands of hours in training a cadre of certified Six Sigma Black Belts (those professionals who can explain its philosophies and principles) and then deploying them to do process improvement work full time, often in a different business unit than the one that hired them. In our experience, most organizations are not in a position to commit the same level of resources, yet they read about GE's success and say, "Let's do what GE did here." This is just one example of the difficulties in taking a copycat approach to the implementation of these tools. Leaders must pay careful attention to the entire set of environmental conditions and patterns that currently exist in their organization and take those patterns explicitly into account as they implement LSS. Setting conditions for a successful LSS implementation is critical for achieving sustainable changes.

Corporate made us do it—*"They made us"* is a time-honored excuse in many organizations for many things. Even though the business environment, staff levels, skill sets, and priorities may be different, "one size fits all" is often imposed on parts of an organization in situations where they may not fit. This comes in handy if (when) the new initiatives go through their rough spots. A reason to abdicate responsibility and accountability is built in. *"We told them it was a bad idea and were told to do it anyway."* Our guess is all of you reading this book have heard some version of that statement, possibly many times before. How motivated were you to implement the new change, or stick with it over time?

Program of the month—*"Let's throw a lot of different programs at the wall and see what sticks."* This statement characterizes some organizations' pattern of reading the latest book or listening to something on NPR (National Public Radio), making a judgment that *"this could be useful to us"* and implementing a program based on the idea. Unfortunately, most organizations don't ever acknowledge the programs that *didn't* stick and don't formally terminate them. Not "pulling the plug" on failed programs makes it very difficult for employees to separate the wheat from the chaff— to know which changes to continue supporting and which to ignore.

It also breeds confusion when similar programs overlap. Larry consults with clients who have implemented the (W. Edwards) Deming PDCA (plan-do-check-act) model, then (Philip B.) Crosby's *Quality Is Free* (1979, McGraw-Hill), then Quality Circles, followed by Total Quality Management, Business Process Reengineering, Lean Enterprise, and Six Sigma. As employees learn cause and effect (Fishbone) diagrams for the seventh time, they reasonably and logically comment, *"Déjà-vu all over again!"* or *"This is the same stuff repackaged in a new container."*

It will impress our customers—A variation on "because it sounds good," organizations in this camp implement process improvement programs because they want to convince customers they care about them. While the words sound great, without a sincere commitment, customers can tell right away that the company is only paying lip service to this concept. This sentiment is captured perfectly in Emerson's quote: *"What you do speaks so loud that I cannot hear what you say."*[*]

---

[*] Quote attributed to Ralph Waldo Emerson. www.quotationspage.com (accessed October 17, 2008).

Leading this type of effort will advance personal agendas—This is a more subtle, but no less toxic reason that Lean, Six Sigma, and other improvement processes fail. Problems occur when *"What's in it for me?"* trumps *"What's the benefit to the organization and its customers?"* Creating a new initiative is perceived as the chance to add budget, add headcount, and increase a leader's influence in the organization. A Six Sigma Black Belt has become a valuable addition to resumes, so classes will fill with participants who have no real intention of applying the skills over and above what is needed to get the cherished Certificate of Completion. In a "refresher" Six Sigma session that one of the authors conducted for a client, a certified Black Belt chose *not* to raise his hand when the group was asked who had Six Sigma experience. You might not be surprised to know this individual—and others like him—had not used their Six Sigma skills for years. Most of the concepts and tools that the company had invested a lot of money in to make available to the organization were forgotten.

"Instant gratification"—Six Sigma, in particular, depends on investing the time and energy to gain a shared, data-based understanding of the current state of a process in order to determine the root causes of the issue under investigation. It is addressing these root causes that enables improvement at the process level. Gathering this data in the "Define" and "Measure" phases of the process, especially when quality data do not already exist (which happens more times than you might think), can take weeks or even months. Organizations that historically move quickly (asap is the fourth highest priority for one client), find such waiting unbearable. We hear leaders ask team members questions like: *"Is the problem fixed yet?" "What do you mean you're still collecting data?" "Do you really need all that information?"* These are asked in tones of voice that make it clear what their opinion is of the Six Sigma process. As you might imagine, the data collection steps are often ignored or compromised, leading to poor solutions (or none at all) and Six Sigma being perceived as a good reason to be yelled at by someone in authority.

OTHER COMPLICATING FACTORS

Here are some other complicating factors that we have witnessed:

- **Leaders often lack an adequate understanding of process improvement tools and implementation issues**. In our experience, it seems especially difficult for people in positions of authority to admit they don't know something, even when they don't. When asked if they are familiar with Lean or Six Sigma concepts, the answer is: *"Of course, we're the 'head honchos,' we know this stuff."* This is delivered in a tone of voice that shouts: *"How dare you even ask the question?"*
- The reality is that leaders often *don't* know this stuff. When project team members tell them the project is going to take longer because the data isn't acceptable, the leader blows a gasket. *"It's always been good enough before. What's the matter with you? Use the information and get on with the problem solving!"* They are not familiar with the concept of Measurement Systems Analysis (MSA), so they react defensively. One of the first questions we hear when coaching Six Sigma teams is: *"Did our leaders attend this course?"* Unfortunately, we often have to tell them "no."
- **When all you have is a hammer, everything looks like a nail**. In the upcoming chapter on Six Sigma, we will explain why it works best with linear, mechanical processes. The underlying premise is that processes have an optimal state; they break for a reason. Once the reasons are identified, these causes can be corrected to return the process to its optimum state. A machine fulfills those conditions; *people do not.*
- Lean, and especially Six Sigma, are implemented with the unquestioned assumption that human systems operate just like mechanical systems. Leaders believe that human systems, like mechanical systems, once fixed, stay fixed. Change the oil, lube the gears, and the system will continue to produce consistent results over time. We argue that this is not true of people. They change, evolve, and adapt to their environments, often in unpredictable ways.

There is one additional complicating factor that gets in the way of successful process improvement implementation and sustainability—traditional change models. Many popular change theories are guilty of the same: "Follow these five steps and you will have a successful change" linear thinking process. Both authors have seen some type of change model taught almost as an afterthought during LSS training. Discussing the messy dynamics of the change process was not given much time or attention. It was more critical to get back to the real tools of data manipulation and DMAIC (define-measure-analyze-improve-control).

HUMAN SYSTEMS ARE DIFFERENT

The rest of this book is based on the premise that human systems are *not* like mechanical systems. They do not respond the way machines do, don't get "fixed" the way machines do, and are not interchangeable the way a machine's parts are. Different models are needed to address human dynamics.

We will propose a different change paradigm that comes from an emergent body of knowledge called Human Systems Dynamics (HSD). It is a framework derived from several scientific disciplines including deterministic chaos, complexity theory, catastrophe theory, and self-organizing criticality. We will introduce HSD in Chapter 4, focusing on the differences that make a difference between HSD and conventional models. We will show how the addition and integration of HSD constructs and tools provide a powerful missing link to improved return on investment in PI and other organizational improvement efforts both in their initial implementation and in their ongoing sustainability.

OUR "SIMPLE RULES" IN WRITING THIS BOOK

There is one other topic we'd like to share in the name of full disclosure. It may be valuable for you to be aware of some additional guiding principles we used in writing this book. We call them simple rules. This HSD tool will be described in more detail later, but for now think of these rules as guidelines for action that provide the framework for our approach. To the extent we have identified and used the rules properly, you should be able to stop at any point in the book and see evidence of these principles at work.

Our Simple Rules in writing this book:

Assume positive intent—Unless people are psychotic, they do what they do for a reason that makes sense to them at the time—maybe not to you or me, but definitely to them. Therefore, the task is to understand their decisions and behaviors in their context, not ours. This shows up in quotes, such as "no villains" or "no naughty or nice."

Accept responsibility—Each of us needs to take responsibility for our own actions, careers, companies—our own lives. Short of being restrained by brute force, each of us makes the decision to continue working at our employer. *"They made me ..."* is only true to an extent. *"We allowed them to ..."* provides greater empowerment, more choices, and more say in our work and our lives.

Explore many paths to success—No two human systems are identical. We believe they are always changing—a little or a lot, for better or for worse, in ways we intend or ways we don't. The consequence of that belief is that no tool or approach is always right. We have already argued that one of the failures of methods like Six Sigma is they have been implemented as the "one tool fixes all." You'll hear (read) the word *context* often in our story. Its importance cannot be over-emphasized.

Tell stories—Stephen Denning writes about the power of storytelling. In fact, the title of one his books is *The Springboard: How Storytelling Ignites Action in Knowledge-Era Organizations.** He makes the case that stories are powerful because readers step into them and fill in the blanks between the author's words with their own experience. When we write, *"Imagine you are working with the world's worst manager ...,"* each of us creates our own definition, remembering our own past experiences of a worst manager. It doesn't matter whether my definition and yours are the same, the words evoke a powerful emotion in both of us. We'll share our stories of real client situations we encountered (names changed, of course) with the expectation that you will fill in the blanks to make our experiences relevant to your situation.

Having established a context for our story, in the next chapters we will describe the historical and current state of two of the most recent, popular, and powerful continuous improvement methodologies: Lean Enterprise and Six Sigma.

---

* Denning, S. The Springboard: *How Storytelling Ignites Action in Knowledge-Era Organizations.* 2001. Oxford, U.K.: Butterworth-Heinemann.

2

Lean Enterprise

It might be easier to say what this chapter of our story is *not*. It is not a comprehensive treatise on Lean history, philosophy, tools, and techniques. There are dozens of entire books whose sole purpose is to accomplish these objectives. Instead, our goal is to provide you with our "snapshot" of Lean, intending that as we discuss the subject throughout the book you will understand our definition of the concept.

A definition of Lean seems to be in order to get started. The definition Larry Solow uses in his work is: "Lean is a business approach that reduces the resources and time required to complete any process by eliminating waste and making value flow to the customer at the speed they require." Said differently, "Use of Lean principles allows products and services to be delivered with high quality, low cost, and rapid response (good, fast, and low cost), all at the same time."

The key point made in discussing the second half of the definition is the key to beating the trade-off, *"You can have it good, fast, or low cost; pick any two"* is simplicity. Fewer steps mean fewer chances to screw up, less time moving between operations, and lower costs due to fewer parts, setups, and tooling.

Brenda Fake likes to compare this to too many cooks in the kitchen, too many hand-offs, and too many leaders. The end product suffers, thus a bad meal.

EVOLUTION OF LEAN ENTERPRISE

The following "history on the back of a napkin" is not intended to be complete, but rather to show that the foundations of Lean thinking have been around for a long time. In fact, one starting point for Lean concepts was 1908. While he didn't call it Lean, Henry Ford identified a faster and less expensive way to build a car. Ford figured that teaching a lot of people to do one thing well produced better results than asking one worker to master a number of different tasks. He kept it simple for a person to learn to do one job well. His Model T assembly line concept reduced the time to build a car by approximately 50% while increasing the quality of the final product.

Henry Ford's later innovations were based on his study of Frederick Taylor and "scientific management." Taylor adapted principles based on the scientific method to identify the critical variables needed to successfully complete a task. Through proper sequencing of those variables and tasks, different workers could be trained to perform any given task consistently, the "one best way."

In the world of Lean, this would come to be called *standardized work*. Using time and motion studies, Taylor looked for ways to get the critical (value-added) tasks of any job performed with the least amount of time and effort possible.

The result of Taylor's scientific method was one of the entry points into another major precursor of today's Lean Enterprise, the work of Dr. W. Edwards Deming.

Dr. Deming's philosophy has been summarized as:

> Dr. W. Edwards Deming taught that by adopting appropriate principles of management, organizations can increase quality and simultaneously reduce costs (by reducing waste, rework, staff attrition and litigation while increasing customer loyalty). The key is to practice continual improvement and think of manufacturing as a system, not as bits and pieces.*

Remember Larry's working definition of Lean as "allows products and services to be delivered with high quality, low cost, and rapid response, all at the same time?"

---

* Cited in Dr. Deming's "Management Training Overview": http://www.dharma-haven.org/five-havens/deming.htm (first accessed October 24, 2008).

Deming provided both philosophy and tools to add analytical rigor to the concept of continual improvement, a concept that morphed into the Japanese concept of Kaizen—many small improvements for the better.

Deming's concepts were utilized extensively in Japan following World War II, and were distilled by Shigeo Shingo, Taichi Ohno, and Eiji Toyoda into what would eventually be called the Toyota Production System, or TPS. The TPS highlighted the distinction between value-added and non-value added activities. Value-added work is defined by the customer and, in its most rigorous form, must pass three tests:

1. Does the customer care if the activity was performed? Does it make the product or service more functional to them? This is sometimes described as: "Would they pay you to perform that task?"
2. Does it physically transform the object of the activity? Adding new information to a form transforms it. Moving the form from the in box to the out box does not.
3. Was it done right the first time? Customers don't expect to pay for rework or scrap.

If an activity is not value-added, what is it? It could be in a category called "business value-added" by some. These activities may be required by law or regulation, and therefore necessary, even if the customer doesn't care. The firm paying its income taxes is an example. Others may be considered as essential. For example, in baseball, the pitcher throwing a pitch to the catcher adds value to those watching the game. The catcher throwing the ball back to the pitcher may not be as valuable to the spectator, but the pitcher needs to get the ball back before it can be thrown again.

Another category of activities falls under the category of nonvalue added work. These are activities that are neither customer nor business value-added. Shigeo Shingo clustered these into seven forms of *muda*, or waste. These wastes include:

1. **Transportation:** The distance work travels between operations.
2. **Inventory:** The piles of raw, work-in-process, or completed work.
3. **Motion:** Unnecessary or inefficient human motion.
4. **Waiting:** Items waiting to be worked on in a batch or queue.
5. **Overproduction:** Building or creating more output than is needed by the customer at the moment.

6. **O**verprocessing: Performing operations or activities that are either not valued by the customer or are more difficult than they need to be.
7. **D**efects: Performing work in such a way that it requires scrap or rework; work that does not meet the customer's requirements.

(These words vary slightly from Shingo's original seven terms. The minor changes and the sequence in which the first letter of each word is presented combine to form a name, TIM WOOD, *an easy way to remember them.)*

The founders of the Toyota Production System created a philosophy and set of tools designed to continuously engage all levels of the organization in an ongoing quest to increase the percentage of value-added work and to reduce everything else. There are dozens of books available that detail the wide variety of tools and techniques available. We identify a few of our favorites in the recommended reading in the back of the book.

Summarizing our history lesson so far:

- Henry Ford contributed the idea of simplifying work so a person can do one thing well.
- Frederick Taylor added "scientific management," "time and motion" studies, and the concept of the "one best way."
- W. Edwards Deming contributed "continual improvement" and quality systems.
- The Toyota Production System provided specific language about ways to identify value, inconsistency, and waste.

In 1996, James Womack and Daniel Jones published *Lean Thinking: Banish Waste and Create Wealth in Your Corporation.** In their book, they synthesized all of the above to identify five principles that are the core of Lean Thinking:

> After interactions with many audiences and considerable reflection, we concluded that lean thinking can be summarized in five principles: precisely specify value by specific product, identify the value stream for each product, make value flow without interruptions, let the customer pull value from the producer, and pursue perfection.

---

* Womack, James and Daniel Jones, *Lean Thinking: Banish Waste and Create Wealth in Your Corporation.* 1996. New York: Simon and Schuster, p. 10.

Each of these principles spawns a variety of tools and methodologies. While describing them here is not relevant for our story, a good introduction to Lean tools can be found in *The Lean Enterprise Memory Jogger: Create Value and Eliminate Waste Throughout Your Company.**

A TYPICAL LEAN IMPLEMENTATION

Having provided some definition and historical context, we turn next to a similar description of how Lean is typically implemented in an organization. While there are many options available, a typical approach involves the following:

1. **Executive overview**: Teach the leadership of the organization what Lean is about and get their commitment to support it.
2. **Create a steering committee**: Often chaired by the vice president of Quality, this cross-functional group of managers and leaders is tasked with designing and managing the tactical implementation of Lean.
3. **Obtain a consultant**: Organizations beginning their Lean journey often invest in outside support to provide consultation, initial training, and project facilitation to jump start the initiative.
4. **Get "quick wins"**: Implement "5S" (a five-step visual organization process named from five Japanese words that begin with the letter S) or other Lean projects that can be implemented quickly and demonstrate a visible difference. These are usually "low investment, low return" projects.
5. **Create current-state "value stream maps"**: Focusing on critical, large opportunity processes, evaluate them using Lean diagnostic tools to identify high priority improvement opportunities. From this, a future-state value stream map and an action plan are developed identifying how to get from current state to future state.
6. **Conduct "Kaizen" events**: Selecting one of the bite-sized areas for improvement identified from step five (above), identify a cross-functional team to blitz the area using relevant Lean tools in a compressed

---

* GOAL/QPC, *The Lean Enterprise Memory Jogger: Create Value and Eliminate Waste throughout Your Company.* 2002. Salem, NH: GOAL/QPC.

period of time; typically two to four days. With the goal of "better, not best," Kaizen teams use their combined wisdom to make immediate changes that improve the targeted area of the process.

MORE ADVANCED LEAN IMPLEMENTATIONS

Many companies choose to stop at this point. They feel good about doing something, involving their employees, and seeing some improvements. Other organizations continue along the path described by Womack and Jones earlier in this chapter. The next steps often require greater amounts of change to the processes and culture of the organization as they:

1. **Create TAKT* times**: Roughly calculated as the amount of product the customer wants divided by the available hours to produce it, TAKT time provides a reference point, a cadence, for creating balanced process flows. This can require significant changes for organizations used to creating batches of work that often flow in unbalanced ways through the process.
2. **Create "pull" and "one piece flow" systems**: Rather than building finished products and storing them until a customer wants them, pull systems trigger work to begin only after the customer places an order. Dell™ Computer's original business model is a great example. A customer visited Dell's website or called an 800 number and spoke to a Dell representative who helped the customer design exactly the computer they desired. They paid by credit card, and then Dell's "just in time" manufacturing process built the specific computer specified by the customer. They accomplished this in less than 30 minutes, requiring about two to four hours of raw materials inventory in the factory. With typical turnaround times of three to five days (including burn-in time) plus shipping, Dell not only delighted customers but had the use of the customer's cash for three weeks until their bill from the component suppliers came due.

---

* The maximum time per unit allowed to produce a product in order to meet demand. It is derived from the German word *Taktzeit*, which translates to "cycle time."

This new business model was a game changer for the PC industry, difficult to replicate in part because of the significant investment already made in existing "push" manufacturing processes.

SUSTAINING A LEAN IMPLEMENTATION

The final element of our "cook's tour" of Lean has to do with sustaining Lean principles and practices over time. Until just before this book was published, Toyota would have been cited as the company that did the best job of integrating Lean into daily business practices. With millions of cars recalled around the globe for acceleration and steering issues, Toyota has come under fire for becoming too enamored of its patterns and being unwilling or unable to look beyond its successes to early indicators of problems with its vehicles. There are relatively few companies who are cited sustaining this over time particularly well. Two organizations who have been named as examples of sustaining Lean over time are LanTech and Goodrich Aerostructures. While they claim successful Lean implementations on their respective websites, neither provides insight as to how they accomplished their long-term Lean success.

Summarizing this chapter, implementing Lean philosophy and tools provides powerful advantages for organizations. Who wouldn't want good, fast, and low cost all at the same time? Typical implementations start small and often plateau at the Kaizen level. Others progress, only to deteriorate over time because they are not enmeshed with other aspects of the business.

As we'll see in the next chapter, similar patterns hold true for Six Sigma.

3

Six Sigma

Six Sigma®* has many areas of overlap with Lean. The term *Six Sigma* is used in lots of different ways depending on context. *Sigma* (which is the lower case letter sigma (σ) in the Greek alphabet), is also the mathematical symbol for a standard deviation. For our purposes, consider it as a very aggressive target for customer-centered quality; the "pot at the end of the rainbow." How challenging? How about executing a process correctly the first time 99.9997% of the time? Some describe this level of performance as "approaching perfection."

Having provided a low-math, snapshot definition, following is a brief description of the history of Six Sigma.

HISTORY OF SIX SIGMA

Since the 1920s the word *sigma* has been used by mathematicians and engineers as a symbol for a unit of measurement in product quality variation. (Note it's sigma with a small s because, in this context, sigma is a generic unit of measurement.)

In the mid-1980s, engineers at Motorola, Inc. in the United States used Six Sigma as an informal name for an in-house initiative for reducing defects in production processes, because it represented a suitably high level of quality. (Note: Here it's Sigma with a capital S because, in this context, Six Sigma is a "branded" name for Motorola's initiative).

---

* Six Sigma is a registered trademark and service mark of Motorola, Inc .

Certain engineers (there are varying opinions as to whether the very first was Bill Smith or Mikal Harry) felt that measuring defects in terms of thousands was an insufficiently rigorous standard. Hence, they increased the measurement scale to parts per million (ppm), described as "defects per million," which prompted the use of the six sigma terminology and adoption of the capitalized Six Sigma branded name, given that six sigma was deemed to equate to 3.4 parts (or defects) per million.

In the late 1980s, following the success of the above initiative, Motorola extended the Six Sigma methods to its critical business processes and, significantly, Six Sigma became a formalized in-house "branded" name for a performance improvement methodology, i.e., beyond purely defect reduction, at Motorola Inc.

In 1991, Motorola certified its first Black Belt Six Sigma experts, which indicates the beginnings of the formalization of the accredited training of Six Sigma methods. In 1991 also, AlliedSignal, (a large avionics company that merged with Honeywell in 1999) adopted the Six Sigma methods and claimed significant improvements and cost savings within six months. It seems that AlliedSignal's new CEO, Lawrence Bossidy, learned of Motorola's work with Six Sigma and approached Motorola's CEO, Bob Galvin, to learn how it could be used at AlliedSignal.

In 1995, General Electric's CEO, Jack Welch, (Welch knew Bossidy because Bossidy once worked for Welch at GE and Welch was impressed by Bossidy's achievements using Six Sigma) decided to implement Six Sigma in GE and, by 1998, GE claimed that Six Sigma had generated over three-quarters of a billion dollars of cost savings*.

By the mid-1990's, Six Sigma had developed into a transferable "branded" corporate management initiative and methodology, notably in General Electric and other large manufacturing corporations, but also in organizations outside the manufacturing sector.

By the year 2000, Six Sigma was effectively established as an industry in its own right, involving the training, consultancy, and implementation of Six Sigma methodology in all sorts of organizations around the world.

That is to say, in a little over 10 years, Six Sigma quickly became not only a hugely popular methodology used by many corporations for quality and process improvement, Six Sigma also became the subject of many

---

* Eckes, George, *The Six Sigma Revolution: How General Electric and Others Turned Process into Profits.* 2001. New York: John Wiley & Sons.

and various training and consultancy products and services around which developed many Six Sigma support organizations.*

IMPLEMENTING SIX SIGMA

A typical Six Sigma implementation is much like an organization's introduction to Lean. In fact, the first three steps are often identical:

1. **Executive overview**: Teach the leadership of the organization what Six Sigma is about and get their commitment to support it.
2. **Create a steering committee**: Often chaired by the vice president of Quality, this cross-functional group of managers and leaders is tasked with designing and managing the tactical implementation of Six Sigma.
3. **Obtain a consultant**: Organizations beginning their Six Sigma journey often invest in outside support to provide consultation, initial training, and project facilitation to "jump start" the initiative.

At this point, Six Sigma implementations often take very different paths.

4. **Conduct a project selection/prioritization process**: The steering committee looks at the needs of the business to identify potential projects. Because there is a major investment of time and money required in the training of "Belts," projects with hard dollar savings (usually >$100 K) are required to obtain a return on that financial investment.
5. **Train Black Belts (BBs)**: For the typical organization wanting to eventually be self-sustaining in their Six Sigma practices, having internal resources with a "wide and deep" knowledge of Six Sigma concepts and (especially statistical) tools is essential. Black Belts (a term taken from the martial arts) undergo rigorous classroom training that can run between 125 to 250 hours in length.

* Six Sigma training, history, definitions—six sigma and quality management glossary. http://www. businessballs.com/sixsigma.htm (first accessed October 27, 2008).

6. **Green Belt training:** This often occurs concurrently, so the organization has skilled resources to support the BBs in their work. Classroom training for Green Belts (GBs) can last between 24 and 80 hours, and is less intensive in its scope and depth.

7. **Conduct initial projects**: Based on the initial projects identified in Step four above, BBs and GBs complete a project as part of their Belt training. Often supported/coached/facilitated by outside consultants, these projects provide a "real world" platform in which to translate classroom theory into practice.

REASONS SIX SIGMA IMPLEMENTATIONS ARE NOT SUSTAINED

Given the significant investment made to this point, organizations certainly intend for those efforts to be sustained over time; to embed Six Sigma into their culture. Unfortunately this doesn't often happen. Here are some of the reasons we believe Six Sigma (and other continual improvement/process improvement) initiatives don't stick.

- **Not explicitly tied into the organization's long-term strategy:** Given the resource-intensive nature of Six Sigma projects, they are easy to dismiss as "a program that has run its course."
- **Lack of clarity about revised roles and responsibilities**: Successful PI projects result in changed processes. While these look great on paper, they often trigger strong resistance among those impacted by the changes. If the new processes are not reflected in updated job descriptions, procedures, roles, responsibilities, and behaviors, conflict can ensue with the organization's own existing written procedures. This results in confusion, bad feelings, and a loss of credibility and trust for the overall PI approach. *"Instead of making things better, it made them worse!"*
- **The folly of rewarding A while hoping for B**: Steve Kerr, the ex-Chief Learning Officer at General Electric, wrote a classic *Academy of Management Journal* article entitled, "On the Folly of Rewarding

A, While Hoping for B."* His premise was when what is requested is different than what is being measured, confusion will occur. Particularly in the case of Lean thinking, "Only build the quantity required to meet the customer's needs *right now,"* often conflicts with traditional measures of efficiency (keep the workers and equipment busy).

OTHER BUSINESS IMPROVEMENT MODELS

While we focus most of our attention on Lean and Six Sigma in this book, these are not the only process improvement methods, models, and tools utilized by organizations trying to improve. A partial list of the methods the authors have learned or taught include:

- Total Quality Management (TQM)
- PDCA (Plan–Do–Check–Act) (Dr. J. Edwards Deming)
- 3^2 Problem Solving (Harley-Davidson)
- P^3 Model (York International)
- Kepner Tregoe
- 8D (Ford Motor Company)

An Internet search will provide additional information on these six models.

We mention these additional process improvement models here because the remainder of our story applies to these every bit as much as it does to Lean and Six Sigma. As you will learn in the "So What?" section of this book, the fundamental issue—and opportunity—has to do with implementing these linear improvement models in nonlinear, complex human systems.

In the next chapter, we will turn our attention to the third construct critical to understanding our story: Human Systems Dynamics.

---

* Kerr, S. 1975, December. On the folly of rewarding A, while hoping for B. *The Academy of Management Journal* 18 (4), 769–783.

4

Human Systems Dynamics

A BRIEF TIMELINE OF LEADERSHIP AND ORGANIZATIONAL DEVELOPMENT THEORY

In the prior two chapters, we provided a brief history of the ways the Industrial Revolution evolved to build things faster, cheaper, and better. In this part of our story, we will focus on a parallel development—theories of leadership and organization. Between the nineteenth and early twentieth century, major changes in agriculture, manufacturing, mining, and transport had an enormous effect on the socioeconomic and cultural conditions during that time. Large numbers of people shifted from being individual farmers and entrepreneurs to working in factories.

The rules were changing. The critical emerging question at the time was: *How can we make more widgets faster and cheaper?* The human condition in the workplace was not a burning concern. The focus was on productivity and mechanization. Failure of managers at the early stages of the industrial age to understand the relationship between employee and machines resulted in a significant disconnect in the work environment. It is not a coincidence that the emergence of trade unions occurred at this time.

The conflict arose between a business wanting to make more parts and the workforce who provides the labor to make the parts. To protect themselves, employees created and joined unions, which can be defined as a "continuous association of wage earners for the purpose of maintaining or improving the conditions of their employment" as outlined in "The

History of Trade Unionism" (1907, *Journal of Political Economy*) by Sidney and Beatrice Webb.*

As we did in the prior chapters, we provide a brief historical timeline of some of the most influential researchers and authors from various fields who addressed the relationship between man and machine in the workplace. While not complete, this chronology gives a sense of the importance of leadership, group dynamics, and differences that emerged at the end of the Industrial Revolution to offset the sameness of the limited management approaches that had developed.

- Henry Fayol wrote about leadership command and span of control (1890–1900).
- Kurt Lewin created new theory in the area of job satisfaction (1920).
- Douglas McGregor introduced "Theory X" and "Theory Y" (1920–1930s).
- The Hawthorne Studies examined the impact on productivity and motivation as a result of perceived interest by management to employees (1924–1932).
- Carl Jung created a taxonomy for identifying and understanding different personality types (1930s).
- Kathryn Briggs and Isabel Briggs Myers developed Jung's theory further to help people understand their preference for work and to make choices (1940s).
- The Tavistock Institute led some of the earliest work in group dynamics by synthesizing research from different disciplines to discover ways to apply psychoanalytic (Freud) and open systems concepts to group and organizational environments (1940s).
- Abraham Maslow introduced his hierarchy of needs concept in a paper titled, "A Theory of Human Motivation"† (1940s).
- Peter Drucker introduced the concepts of decentralization and simplification, Managing by Objectives (MBOs), and the knowledge worker (1950s).
- Warren Bennis provided new concepts of group behavior and democratic leadership (1960s).

---

* First published in 1894, it is a detailed and influential accounting of the roots and development of the British trade union movement. The research materials collected by the Webbs form the Webb Collection at the London School of Economics
† Maslow, A.H., A theory of human motivation, *Psychological Review* 50(4) (1943): 370–96.

- Christopher Argyris introduced learning organizations theory and action (1970s).
- Fred Trist and Edward Emery identified sociotechnical systems to speak to the interconnectness of *social, technical,* and *environmental* aspects of an organization (1960s–1970s).
- The Tavistock Institute also created some of the first structured thinking around Self Managed Work Teams: self-organized, semiautonomous, small groups where members determine, plan, and manage their day-to-day activities and duties (1980s).

All of these approaches focused on the human side of change. Sometimes psychological and intellectual capacities of individuals were central to the story. At other times, the group or social unit drew more attention, but always the focus was on the people and their relationships as different in kind from the forces and features of the world around them.

While analysis and action for material and technical systems moved forward along one path, theories about human intelligence (and intelligence about humans) moved along a very different path. It was not until the 1980s that developments in theory and demands in practice generated a new approach to Human Systems Dynamics (HSD).

DIFFERING ASSUMPTIONS

But before we delve more deeply into HSD, three basic assumptions about people and their worlds demanded this stark distinction between human and material systems approaches. First, human systems are open to influence from many different directions, all at the same time, while material ones could be isolated from extraneous forces. A car will never misbehave because it had a fight with its spouse the night before or be distracted by memories of a traumatic experience decades before. People, on the other hand, are prone to such distractions.

Secondly, a very large number of factors influence how one individual behaves, and the number of relevant variables is astronomical when two or more people are involved. Physical systems, on the other hand, are driven by three variables: mass, distance, and time. Virtually everything you might want to say about a physical system can be described with some combination of these three things. Temperature, for example, is the distance a column of mercury is pushed along a tube. Force is equal to mass times acceleration and acceleration is equal to distance divided by time squared. Though there have been many efforts to define the fundamental variables that shaped human behavior, none was applicable in all places and at all times.

Thirdly, human beings interact. Each one influences the other in many different ways and all the time, while each is also influenced by others. This mutual relationship, called nonlinear causality, is a fact of life in human systems, everywhere, all the time. Though physical systems also can engage in nonlinear causality (consider a thermostat or a hurricane), their behavior can usually be boiled down to single, one-way causal links.

Thus, the technical systems approaches emerged to take care of closed and linear causality and the social systems approaches emerged to take care of open and nonlinear causality. Two quite distinct languages and methodologies emerged and conversations between the two were few and far between.

We describe these two parallel paths of system/process development and people/organizational development because the intersection of these paths provides the backdrop to what we believe is the next groundbreaking work of management and organizational theory: Human Systems Dynamics.

We've mentioned HSD in bits and pieces; almost as "coming attractions" throughout our story so far. We now introduce HSD much the same way we have Lean and Six Sigma. Rather than provide an exhaustive textbook treatment, our focus will be on what sets it apart from traditional approaches to change management for business process improvement.

In the 1970s and 1980s, physical scientists recognized that some of their most important systems were open and nonlinear. Weather, lasers, ecology, thermo- and hydrodynamics, and heart rate were some of the systems that challenged the traditional assumptions of physical and technical systems. Though they were material in nature, they didn't always behave in predictable and rational ways. To meet this challenge, scientists and mathematicians explored new ways to understand, describe, and manipulate what they called the chaotic or complex behaviors of these strange and surprising collections of physical stuff. The result was a broad-based movement that created a wide variety of theories, methods, models, and tools for understanding and influencing how physical systems behaved. Though the movement crossed traditional disciplinary lines, the whole is often referred to as "nonlinear dynamics."

When physical systems were recognized as open, multivariable, and nonlinear, they began to approach the behaviors of humans in their inter- and intrapersonal interactions. Some social scientists recognized the power of these new approaches and applied them to improve the understanding and action of people and organizations. One strand of that inquiry came to be known as Human Systems Dynamics.

A short, concise definition of HSD is: "An emerging field of research and practice that takes into account the principles of complexity, nonlinear dynamics, and chaos theory as it applies to the study of groups as they live and work in teams, organizations, and communities."*

HSD proposes that human systems (e.g., organizations) emerge as a result of a series of complex interactions between various individuals and groups within the whole. Agents (e.g., the marketing department) interact (or not) with the operations function to help support the reason they are all (at least theoretically) connected in the first place (running the business).

> "Based on complex adaptive systems theory (Figure 4.1),[†] HSD goes beyond the traditional, linear models of systems that rely on step-by-step planning and expectations. Linear models assume predictable environments and predictable outcomes with high levels of agreement and certainty. But such linear models of thinking and functioning break down in the unpredictable, emergent environments of today's social and economic systems. In contrast, HSD is an innovative way to think about organizations. It recognizes that most human systems evolve from complex interactions, so its descriptions better match reality of behavior and action in individual, group, organizational, and community behavior."[‡]

---

* Human Systems Dynamics Institute Web site, http://www.hsdinstitute.org/about-hsd/what-is-hsd/faq-about-the-field-of-hsd.html (accessed October 28, 2009).

† Adapted from HSD at work: Frequently asked questions about human systems dynamics, Human Systems Dynamics Institute Web site; http://www.hsdinstitute.org/learn-more/library/faq-booklet.pdf (Adapted from The Human Systems Dynamics Institute. With permission.)

‡ HSD Institute Web site; http://www.hsdinstitute.org/about-hsd/what-is-hsd/faq-about-the-field-of-hsd.html.

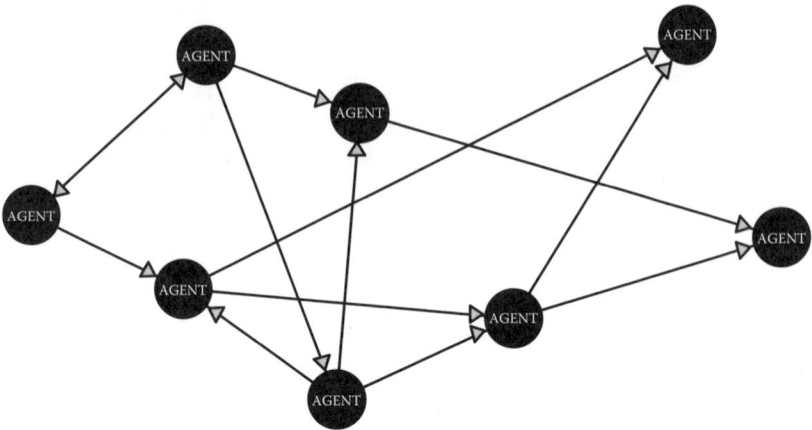

FIGURE 4.1

Complex adaptive system (CAS self-organizing system). (From Human Systems Dynamics. With permission.)

Viewing organizations as complex adaptive systems is a major paradigm shift. It explains why traditional methods for implementing and sustaining process improvement have not worked. HSD offers new tools and new possibilities for action by individuals and organizations. A sample of these tools will be described in the next sections of the book.

The self-organizing, emerging character of complex systems can be seen everywhere. One example is the story of how this book came to be written. As described in the Preface, from a strictly linear, "cause and effect" standpoint, the chances of Brenda Fake and Larry Solow joining forces with Glenda Eoyang and her work in human systems dynamics was astronomically small—yet it happened.

We have provided a basic background to HSD in terms of its theories and how it connects the behavioral and hard sciences. This bridge is the difference HSD can make in implementing the linear approaches of process improvement within the complex human systems that define today's organizations.

Section I Summary

We have written this book to help those professionals who implement process improvement in their organizations to do so more effectively and efficiently, and to increase their ability to create real and lasting culture change. We want readers to understand and apply concepts and tools based on nonlinear dynamics that offer new possibilities for action.

Toward that end, the first section of our book provided a starting point; a description of the context and current states of the systems we are interested in discussing. We have provided our rationale for writing this book as well as short historical backgrounds of Lean Enterprise, Six Sigma, and Human Systems Dynamics (HSD).

CASE STUDY INTERVIEW

We are aware that our history lesson may not be sufficient to fully convey the dynamics of this current state of process improvement. To provide another perspective, we have asked two former colleagues and clients to provide their stories. We were asked to change the names of our interviewees as they both still work for their current organization and expect to continue to do so.

The following stories are provided by two very different, yet equal participants. Each of these leaders has played a role in a large-scale implementation. Six Sigma, in their company, and each has a perspective to share. The conversation that follows is the result of a series of interviews with each of the leaders conducted by Brenda Fake over a period of two weeks in May 2009. In this first part of the conversation, you will learn about the organization's background and the issues these change agents confronted as they tried to transform their company's culture. Later in the book, after we have introduced our new tools and concepts, we will revisit the story to find out how these tools made a difference.

PANEL DISCUSSION: THE CURRENT CONDITION

The organization is a large global, high-tech manufacturer that competes in multiple markets. It sells directly to consumers and is also a major business-to-business supplier. The company is complex, with mergers and acquisitions resulting in multiple core businesses housed in various manufacturing plants and engineering functions located around the world. As a result, a dependency on a virtual environment is emerging.

The organization has two distinct cultures from one of the larger mergers in its history. It has a commitment to high quality and customer service across the organization, but struggles with program planning and pricing custom design services to customers. The firm remains competitive and is expected to be a major player in the markets it serves in the future.

Brenda Fake (BF): Good afternoon. Would you please tell us your name and describe your experience, background and role in quality, process improvement, Six Sigma and/or Lean Enterprise since your company decided to implement Six Sigma?

Ron Fischer, Ph.D. (RF) (not his real name) Six Sigma Master Black Belt: Ok, I'll start. My background has been as an industrial statistician working with engineers and others who need data converted into meaningful information that can help in decision making. I have a Ph.D. in statistics, undergrad degree in math education, and over the years I have been able to build a good reputation as someone who can analyze data to extract meaning.

Markey Keating (MK) (not her real name) Change Leader Specialist: Hi, I am so excited to have this opportunity to discuss Six Sigma with you. It is best for me to start from the beginning and where I was when Six Sigma started. I had just returned from a two-year European assignment and upon my return was offered the opportunity to get the Black Belt training. I was told that completing this process would get my career ticket "punched" and help me get to a director level in the organization. I have always been very conscious about building my career and taking advantage of opportunities to achieve that goal. My background is not only in engineering, but also in sales and marketing, customer service, and competitive intelligence. Additionally, I added the Lean Expert certification to my resume and I am currently working as a change leader specialist for a business improvement division of engineering. Regardless of my role, there has always been a process improvement element (either minor or major percentage) of my job. I have experienced minor and major success in my process improvement activities.

BF: How did Six Sigma start at your company?

RF: I can offer this—from what I recall, a Manufacturing VP from another large high-tech manufacturing company, which had deployed Six Sigma, whispered in the ear of the then current CEO that "$1.6 billion can be saved in your company if you consider implementing Six Sigma." This new VP had assessed our company as currently being a "3-sigma" company. He figured 10 to 30% cost savings could be achieved through Six Sigma reductions in waste and defects. The company was around $8 billion in sales at the time. The CEO at the time was known for being very decisive, but not so good at communicating a vision for the future; his focus was clearly on getting relatively short-term results.

BF: Interesting, but not surprising. So how did the actual deployment unfold if visionary communication from the leader was challenged?

RF: The deployment started with a call for every site to identify people to start the improvements at their respective locations. The request had focused on monitoring the key production metrics. Unfortunately, the first people they sent were people with the wrong skills. They thought it was a metrics exercise to supply the data to corporate on productivity and quality metrics. The consultant hired to coordinate and advise the deployment threw a fit. He wanted manufacturing and quality engineers who could engage in the factory and kick butt. Like a true BLACK BELT. After the definition was understood, the site selected me to be their first "Black Belt." I had gone to Black Belt training a year earlier and was then sent to the first Master Black Belt (MBB) School conducted by the consultants who were hired to train leaders internally so the process of knowledge transfer would begin for the organization. The next step in the deployment was for those who had attended the consultant-driven training to develop the curriculum for internal training of MBBs so the company could develop and transfer knowledge with its own staff.

MK: I didn't have the same line of sight RF had at the time, but what I do clearly recall is that both Six Sigma and Lean rollout processes had different requirements for different groups and varying depth requirements depending on roles. For example, there were different requirements and training for three different areas of the company: Sales and Marketing, Operations, and Engineering. There were at least three levels of training: Foundational three-day class, Black Belt four-week class, and the Master Black Belt eight-week class (included first four-week class plus additional four weeks). At minimum, everyone in these three organizational areas had to take the three-day class and complete a project using some of the new tools and processes. Project selection was coordinated with the director of the department, manager, and employee. The employees' directors were expected and tried to make the projects part of the daily work. One thing I recall from the training experience was how mentor-dependent the training and certification process was. An attendee's experience or perspective was largely based on the relationship or impression of the mentor. If the mentors were too theory-based, the experience usually left a less than favorable impression of Six Sigma with participants, but if the mentor had a practical approach, it usually was a favorable experience. Overall, it seemed there was an inconsistent mentor base and approach to deployment.

RF: Hey, I was one of those mentors! But I, too, understand the perception because of my own experience. At the same time, I was teaching as many as five four-week classes per year and mentoring the projects of around half the students (60 plus). This was a major career shift from statistics to internally consulting in a practical manner to focus on the project for results. Because of my statistical skills, I was leveraged across several projects helping internal customers and other Black Belts and MBBs who needed more advanced stats knowledge. Most of my help was validating their work and to be a guide to help bring the data analysis to a real decision. There was an interesting mindset among some of those in Six Sigma. They tended to believe the objective was to apply statistical tools. It sounds like that was perceived as too theoretical. I basically found turning projects to practical business results nearly a full-time activity.

During this time, the company acquired and merged another company into the larger organization and introduced Six Sigma to the new entity. I was then assigned to be the liaison to engineering as a Master Black Belt. My first assignment was to deploy a product design scorecard. It was clear that a Six Sigma approach for product design was needed so the scorecard would have something to measure. I then worked with others to develop the curriculum for the Design for Six Sigma (DFSS) course and taught several classes for DFSS trainers. The next place to go was Lean Product Design as Master Black Belt and before I knew it, I realized that I now had over 15 years as a Six Sigma expert.

BF: Can you describe the tactical approach to how Six Sigma was implemented for knowledge transfer and business results?

RF: Well the rollout was two-fold: a training rollout coupled with a project rollout. It started with the project plan. The projects were supposed to be chosen to best improve the site's business performance. The people were then chosen from the targeted areas so that people could attend training with a project in hand to learn the tools and application of Six Sigma. This was a learn-by-doing approach. It was a good model that used adult learning methods. In the longer term, additional projects could be selected and the Black Belts would then work them and gain the results even faster because they didn't need the training.

In parallel, there was communication to all the site leaders from the top that touted the benefits of Six Sigma, how Black Belts would be trained to make breakthrough improvements, and how the site champions would be coordinating the efforts at the site. Special communication was added to the annual report for shareholders. The rank and file had items in newsletters that Six Sigma was coming and their participation was needed. The communication roll out tried to hit all the levels of the company and all the stakeholders about the Six Sigma program.

BF: How well did that work for the organization?

MK: Hey, I can start with this one. For early adopters, or those with a positive belief in process improvement, the training and project went well. A "good" mentor helped the new users blend Six Sigma into their current work practices. There seemed to be an increased ability for analysis in nonmanufacturing parts of the company like marketing, and then addressing a need for less detailed and strenuous analysis tools in areas like engineering that could meet the quick turnaround needs of the situation.

RF: That's good to hear because it validates my perspective that the training was very effective in raising the quantitative literacy of the organization, and especially, the skills of the Black Belts in improving processes. Many employees realized they did not have this skill set and sought help from their local Black Belt to achieve improvements they needed. The Black Belts were usually highly visible and the people in the sites knew who could help them with data and analysis.

BF: This is good news. What didn't work so well?

RF: To start, our efforts in managing organizational or process change did not work.

MK: Why do you think my job exists today?

RF: And that's a good thing because in the beginning, projects often identified sound changes that would generate the projected savings, but as soon as the key people moved on, the processes backslid to the old ways. There was no transfer of knowledge, in-depth participation, or leadership stewardship of the changes going on. Next, the project selection did not work as effectively as needed for the business. The projects were completed and savings documented, **but** the projects did not have an enterprise focus and they did not align to a bigger picture.

The training did such a good job of improving people's skills that the Black Belt designation became a coveted perk in the organization. This increased demand for the training and drove poor behavior in project selection. Since one had to have a project to go to class, often projects were selected that lacked alignment with what was needed for the business. The training was great, but the project deployment suffered because we found technical solutions, but we did not find the cultural solutions to take full advantage of the education and intent of the effort.

I really thought from the beginning, when I was designated to go into the first class as a Black Belt, that I would be more effective if I could get some change management skills. I was a statistician and already knew the technical side of the equation, but needed the change management and consulting skills for my work to be effective. But what I got was more technical skills and not the stuff I really needed. If you can't switch from analytical to practical when needed, you can't be effective in this work. Management did not understand the role Six Sigma should play. Often, projects added a Black Belt to show that the team was addressing an important issue and the Black Belt often made significant contributions, but didn't need to use much of their technical skills. Instead, they had to use skills they weren't trained in—the skills of organizational change management.

MK: Well from my perspective, here is what I thought could have been better. First, a segment of the population was not going to try anything new; they were discouraged by the "flavor of the month" initiatives in the past, and this effort was not perceived as different. Depending on the locations, it would vary by what the location would accept or not accept. The flavors of the month tasted like business reengineering, Mark Hammer, TQM—I think you get my point here. Second, the rollout did not emphasize how to adapt the language, tools, and processes to fit with the audience's experiences. In some cases, the new users could not see how to take the concepts and apply them to their work. The process was very dependent on the mentors and, if the mentors did not have an ability to translate from theory to application for a given function, the experience was negative. Another contributing factor to the mentor issue was that one of the larger locations only wanted you to be a Black Belt for 18 months, resulting in a big mix of mentors from various areas of the companies. If the mentor had a strong bias from where he or she had come from—for example, quality audit—that mentor would be more into work on compliance issues because that is what he or she knew. It seemed that everybody experienced the Ph.D. Black Belts in a negative way because they were perceived as not adaptive. Right or wrong, it is the perception I am afraid that still exists today.

RF: Ouch!

The next part of our journey is titled, *"So What?"* The first part of the section introduces a more extensive case story that highlights the business and people dynamics that occur during a typical process improvement implementation. The story is then retold, this time embedding a greater awareness of organizations as complex systems.

Section II

So What?

5

Introduction to "So What?"

In the previous section, we described the fundamentals of Lean Enterprise, Six Sigma, and Human Systems Dynamics (HSD). The next part of our story returns to the reason we felt compelled to write this book in the first place—that utilizing a linear approach to implementing Lean, Six Sigma, and other continuous improvement (CI) processes is fundamentally flawed.

Process improvement methodologies are valuable. Each has powerful tools and logic based on a linear, step-by-step approach. They are founded on the underlying paradigm that organizations and their processes operate "like well-oiled machines."

The problem is that humans—individually and collectively—don't work that way. People and the organizations they belong to are not—cannot—and never will be machines. In our description of HSD, we stated that humans are complex creatures with individual differences, complex social structures, and infinite communication patterns. Systems that survive and flourish are adaptive and address the conditions necessary to sustain themselves. The understanding of these surroundings can be conscious or not, but understanding and addressing the conditions necessary to self-organize is "the difference that makes a difference" for process and/or business improvement efforts.

An analogy might be asking a professional woman to organize luncheons for a group of people in Medicare-funded retirement homes with the objective of fundraising to build her business. The missions are different and the objectives don't align. There is nothing wrong with either mission and nothing to be gained in blaming one or the other for failing to reach their respective objectives. Each is useful in certain situations. The conflict arises when there is pressure to fit the two together.

We hope to make these concepts come alive by sharing a case story. We use the case story language to differentiate it from a case study. A case study is typically dense, factual, and complete. Our case story is intended to highlight, in a dramatic and engaging way, selected dynamics of a typical Lean Six Sigma implementation. Said differently, we designed this tool much like a designed experiment. We chose several variables of interest and set aggressive limits for them. In our case story, we made deliberate decisions to ignore or control other dynamics that were not of interest to us.

This section will provide a landscape for us to describe how one fictitious organization acted as it implemented a process improvement initiative. The company is a composite of dynamics the authors have actually witnessed in client situations. The story then will be repeated in order to emphasize and highlight what is possible when the concepts and tools from HSD are applied.

We will include several players who are involved in process improvement and quality efforts—the corporate officers (the C-Suite*), the team leader, project team members, and others in the organization.

The case could be about any business, in any industry. We might have written about:

- A small hotel chain that acquired a new property from a struggling competitor and is trying to make the new acquisition "work as smoothly as we do."
- An insurance company struggling to keep up with rapidly changing federal and state regulatory changes.
- A small nonprofit organization whose target audience is growing at the same time its funding sources are shrinking.
- A hospital that is trying to reduce the complaints of customers who are waiting too long in an emergency room.

We chose to write our fable about a manufacturing organization. We did this because our fictional company shares some of the dynamics we experienced working with a client with whom both authors have a great deal of experience.

---

* This is the group of chief officers of a business organization, with chief executive officer (CEO), chief operating officer (COO), and chief financial officer (CFO) usually being included, and sometimes others with titles like chief learning officer (CLO) or chief technology officer (CTO). These often vary within and between industries.

Although many of the issues are the same, we recognize there are significant differences in service and nonprofit sectors. While we believe the theories and tools provided are relevant outside of manufacturing, we do not assume manufacturing and other business segments operate exactly the same way. Hopefully, readers with experience in the service and nonprofit sectors will see patterns and dynamics that will apply to their situation.

Again, the forthcoming case story is a dramatization that amplifies the dynamics of a traditional process improvement implementation. This is done for two reasons: (1) to highlight the perceptions such initiatives can create in an effort to improve the business and (2) to set up the second telling of the same case story.

The story of *TryinHard Marine* is related in the next chapter.

6

Introduction to TryinHard Marine

TryinHard Marine (*THM* for short) is headquartered in San Diego, California. It manufactures sophisticated electronic navigation systems for ocean-going civilian and military ships. The San Diego facility houses centralized corporate functions as well as manufactures one of *THM*'s "bread and butter" products, the OWow 123 NavSystem. *THM* is considered a leader in the industry with annual revenues of $500 million, and has approximately 3,500 employees. The San Diego hourly workforce is represented by the International Community of Electrical Workers (ICEW).

THM has four other facilities, which are strategically located in the port cities of Miami, Charlotte, Seattle, and Philadelphia (Figure 6.1).

Each site contains a sales function to service major shipbuilders and military installations. Each also manufactures its own product line. These are smaller facilities, with the number of employees ranging from 350 to 1,200. The Seattle and Philadelphia hourly employees are represented by the ICEW, while the Miami and Charlotte hourly workforces are not affiliated with a union.

TryinHard Marine has three major competitors. *GoGetem* is headquartered in San Francisco and has recently come out with a serious competitive product in the OWow123 product and market space. *BetterThanYou* is a new entrant in the field, competing on price. While they don't have a product that competes directly with the OWow123, they are gaining a foothold in *THM*'s other product lines.

JapanRocks! is the "800-pound gorilla" in Asia. *THM* has not considered them a domestic competitor, ceding the Asian market to them while focusing on the North American and Western European markets. Recently, though, *JapanRocks!* opened sales offices in each of *THM*'s four cities. So far, loyal customers have reported that *JapanRocks!* is making

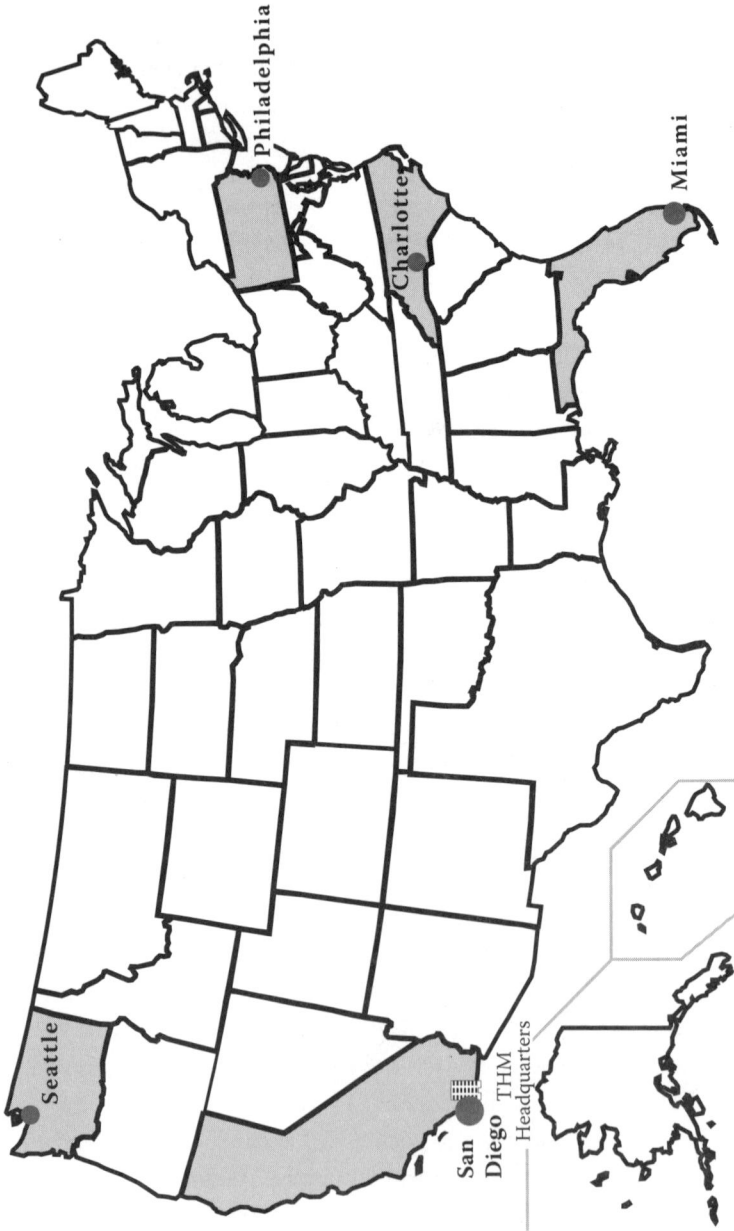

FIGURE 6.1
The plant locations of *TryinHard Marine*.

claims that with today's virtual communication technology (which they purchase for their clients) and declining shipping costs, geographic distance need no longer be a significant factor in making a purchasing decision. With products that rival (and might even exceed) *THM*'s in terms of quality, *JapanRocks!* has become a powerful competitor. Did we mention their prices are on average 12 to 15% lower than *THM*'s?

John Saylor, chief executive officer of *TryinHard Marine*, had been leading the organization for about three months. He had been contacted at that time by *THM*'s chairman of the board, G. E. Welch. Welch told him that *THM* was looking for a leader who could expand the business and aggressively execute a global strategy, while expanding products and services into the current customer base. It was a big move for John, and he was ready to make his mark in the industry.

He had been the Number 2 person at a small, high tech competitor that designed and built unmanned aerial vehicles (UAVs) for the military market. The company was purchased by a larger competitor because their designs were superior to anything in the marketplace. At *THM*, John planned to grow the existing business and begin a strategy of acquisitions to expand *TryinHard Marine*'s global reach.

John was ready to do whatever was necessary to achieve personal and organizational success. He spent long nights studying *THM*'s books and operations. He even joined CEO Roundtable™, a network of like-minded leaders, as part of his professional development to share his business challenges and obtain experienced advice. In return, he would offer his expertise when appropriate. The organization met once a month as a professional support and mentoring group.

John's first step in executing his business strategy was to assess the *THM* organization (Figure 6.2) and build an immediate plan to improve North American and European market share. *JapanRocks!*'s expanded domestic presence was a major concern. He called a staff meeting to discuss this new threat.

Sally Dow, CFO, was in attendance. She had worked her way up the organization over the past 10 years. She started in the accounting department after graduating from college and completing an accounting internship from the University of Minnesota. Sally was respected by the organization for her level approach and grasp of the business. Her father had worked at *THM* in the early years as a lead engineer. She viewed cash flow as the life blood of the organization and used the budget as a guide to influence

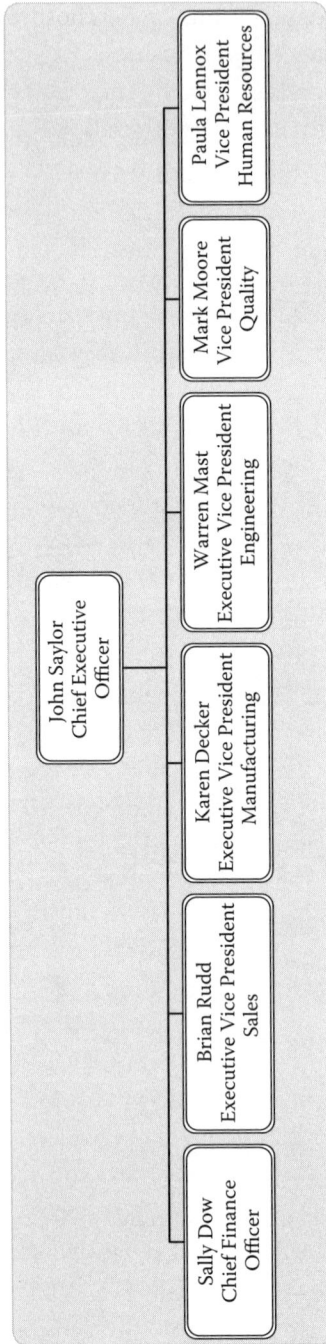

FIGURE 6.2
TryinHard Marine Organization Chart.

THM's decision making. Sally was a straight shooter who did not have a lot of time for esoteric conversations or for those who did not understand the critical role of finance in the business. Extremely organized and dependable, Sally had emerged as a solid finance leader. So far, she enjoyed the new energy John Saylor brought to the company and looked forward to working with him in growing the organization.

Brian Rudd, executive vice president of sales, was an interesting person in that he came from the engineering side of the business five years ago. His friends in engineering would later say he "crossed over to the dark side." Brian's background was in electrical engineering from Penn State. He had a solid understanding of the technical side of the business—necessary for selling *THM*'s complex products. Brian demonstrated early in his career to Warren Mast (his boss at that time) that he had a gift for connecting with the customers and ensuring all the requirements on the designs were something *THM* could actually manufacture. His ability to manage projects and exceed customers' expectations was his strong suit. Warren had encouraged him to go into sales, partly to ensure *THM* was not "making promises with its sales force its design team couldn't keep." Lately, however, with increasingly changing customer requirements, and without direct control over the design and production areas, he was challenged in meeting customer demands and needed a new approach for promoting *THM*.

In John Saylor's initial assessment of his direct reports, he considered Karen Decker, executive vice president of manufacturing, his strongest overall general manager. She had a clear line of sight to all the facilities' production numbers. She was well liked and respected by both the employees and union leaders as a good leader with excellent follow-up skills. Karen had a reputation for getting back to people and was often quoted as saying, "You may not like the answer, but it *is* the answer and we always need to consider the customer's requirements and expectations." Karen had a direct, yet friendly, approach that motivated people to "go the extra mile" for her. She had strong communication skills and all of her team leaders knew their targets and were expected to deliver results. Her biggest challenge was that the sales department would make requests late in the production cycle without changing the original delivery date. These requests created tremendous frustration throughout her organization. Repeated conversations with Brian Rudd didn't seemed to make a difference.

Mark Moore, vice president of quality, was excited about John Saylor's abilities and the emphasis he placed on quality. Mark believed quality was the differentiator in his previous business. After a short stint in a small machine shop, Mark had worked at *THM* for almost 20 years. With a manufacturing engineering degree from the University of California /San Diego, he moved after eight years in manufacturing to operations. Mark enjoyed helping to create the design and then building it to delivery. He was proud of his ability to ensure the designs were something that could actually be built and easily improved upon.

His skills in applying new methods to current work practices and developing standards resulted in him being recruited into the quality organization almost five years ago. Mark was new in his position as vice president, being promoted to that spot by John Saylor in one of his first personnel moves since arriving at *THM*.

Paula Lennox, vice president of human resources (HR), rounded out the meeting. Paula had been with the organization for about four years and started as the HR manager for sales. Her background was in service and sales with two other global manufacturing organizations. Her experience with labor law made her a good choice for *THM*. In addition, Paula enjoyed the generalist role of HR and made an impact in many different areas of the organization. Two of her biggest accomplishments to date were building a strong succession plan and revising the compensation process. She also created a leadership development program, which had been credited with helping specific individuals improve, but the jury was still out about its impact on the larger organization. With some recent attrition in the HR area, Paula found herself challenged to meet the transactional demands of the business while participating in the strategic side of building the organization.

Warren Mast, executive vice president of engineering, was at a client site doing some problem solving and was not present for the meeting. Warren, a 27-year veteran of *THM*, had spent his entire career there since graduating from the Naval Academy with a degree in mechanical engineering and serving his country. He was known to run a tight ship and stood up for his engineers. Warren expected engineering to be involved early on in the design process to influence the decisions that impacted his organization. He was particularly sensitive to complaints that engineering issues caused errors in production. Warren took pride that *THM* had been recognized by its customers for the engineering built into its new designs.

After summarizing the *JapanRocks!* threat, Saylor opened the floor to suggestions about how to best counter the new competitor. The discussion was lively and many ideas were offered, discussed, and rejected. Price cuts would cut too deeply into margins. Greater investment in R&D would take too long to pay off. Creating a new marketing effort was discarded with the statement, "We don't need more sizzle; we need a better steak." Acquiring a small competitor in Asia was considered. After considerable conversation, this idea was filed for long-term consideration, but rejected as a viable short-term option.

Asking leaders to "get tougher and ask more" of the employees was given serious consideration. They agreed it was likely that *THM* suffered from a degree of complacency as a result of being the market share leader in North America. Surely, cracking the whip a little would produce immediate results. Everyone at the table agreed they had been more than fair to the employees. Wasn't it time for them to return the favor?

These discussions had particular urgency because John Saylor had made it very clear that once issues were discussed and decided, he expected alignment and a united front by his staff. No exceptions.

At one point in the conversation Mark Moore mentioned he had heard that their low cost competitor, *BetterThanYou,* was using an improvement program called Lean Six Sigma (LSS) to achieve significant cost savings. Mark reported Lean Six Sigma was a rigorous, customer-focused, data-based process created by Motorola and brought mainstream by such premier companies as GE, AlliedSignal, and others. Moore went on to share that GE reported savings of over $7 billion dollars in annualized savings since its inception in the mid-1990s. Six Sigma utilized internal employees, working in teams, to identify and improve processes. Companies that implemented Lean Enterprise, a complementary process improvement initiative, claimed 40 to 65% decreases in cycle time. Several large consultancies existed that specialized in helping organizations implement both Lean and Six Sigma initiatives.

There was a discernible increase in energy as Mark finished his comments. Lean Six Sigma appeared to be a much more powerful way to counter *JapanRocks!* than yelling at employees. After all, with world-class organizations—ones who manufactured sophisticated electronics by the way—reaping tremendous gains, the chance to improve quality and reduce price seemed assured.

Brian Rudd, EVP of sales, was enthusiastic. His sales reps would have something new to bring to their accounts immediately, i.e. "We're implementing Lean Six Sigma to serve you better," and the results would provide success stories to share in the future.

Karen Decker, EVP of manufacturing, was a little more hesitant. She operated in the world of regulations, implemented by the civilian government as well as the military. Changing processes would be a major hassle. Documents would need to be revised, new qualification testing might be required, and significant amounts of unplanned and unbudgeted time would have to be spent to make these process improvements.

Sally Dow sat up straighter in her chair when the word *unbudgeted* was spoken. She asked Mark what he had learned about the costs and the return on investment (ROI) associated with Six Sigma. The $7 billion of savings was nice, but how much did it cost to get it? Mark had anticipated the question, and he reported the average Six Sigma project returned an ROI of between 1.5 and 5:1. Projects typically lasted from 3 to 6 months and generally broke even within 12 months. Sally relaxed a little. With no significant short-term impact on the balance sheet and a tremendous upside, she assured the rest of the leaders present that she would find a way to get them the start-up funds they needed. Sally added that she wanted to see some additional information on the reported ROI results to which Mark had referred.

Paula Lennox, HR, was quiet throughout. Her only comment was that San Diego *TryinHard* hourly workers were unionized and bound by a labor agreement. Organizational improvement was clearly needed; HR was onboard so long as Six Sigma implementation and projects stayed within the confines of the collective bargaining agreement. She wondered to herself whose budget would be hit for the training, but said nothing.

John Saylor asked for a minute to summarize the conversation and poll his leadership team. Implementing Lean Six Sigma at *TryinHard Marine* seemed like it could be the answer to their problems:

- It would result in higher quality and lower cost, allowing *THM* to compete not only with *JapanRocks!*, but with the growing low-cost threat of *BetterThanYou.*
- LSS offered sales an immediate story to tell to their existing accounts, giving them a reason to ask for a meeting and get some face time.

- While implementation would cause short-term stress in manufacturing, Karen would eventually be the beneficiary of better quality and more controlled processes. It certainly seemed like a better solution than supervisors jeopardizing relationships with valued employees by demanding more work output.
- Sally seemed okay with the financial aspects. *THM* would go out of pocket for a short time, but the first wave of projects should begin generating a positive ROI in three to six months (probably closer to three; *THM* not being as big or administrative as GE). Those savings would fund the further rollout of the initiative.
- By signing up with a reputable consulting firm, they would have a partner that had already been through the LSS learning curve, so their implementation should be relatively smooth. After all, *THM* was nowhere near as large as the behemoth companies Mark cited as success stories.

Bonus Thoughts: In a recent CEO Roundtable meeting, other CEOs had been starting to wonder out loud if Six Sigma was right for their business. Implementing Six Sigma at *THM* provided an opportunity to gain some bragging rights and move past talking and into action.

After sharing these thoughts aloud with his team, he asked, "What are we missing? Every day we delay, our competitors are gaining ground on us. Are we ready to actively pursue Lean Six Sigma?"

Brian said, "Warren Mast isn't here. Does anyone think he would have a problem?" Without hesitation, everyone said no. "This is a data-driven process and engineering is a data-driven discipline. Warren is always pushing us to get ahead of the customers, and it sounds like Six Sigma focuses on customers."

John thought that Warren still deserved to state his own perspective. Paula agreed and left the meeting to contact Warren's assistant to see if he could be reached to conference into the meeting by phone. Unfortunately, Warren was on a flight to the Florida facility and not available.

Brian then commented, "What part of LSS wouldn't Warren like?" As heads nodded in agreement, John asked for a formal vote, "All in favor of moving ahead immediately to aggressively investigate implementing Lean Six Sigma at *THM* raise your hand." Every hand went up without hesitation.

"I guess we have a decision," said John. "Mark, since Lean Six Sigma appears to primarily be a quality improvement program, I'd like you to lead this effort. Okay?" Mark nodded affirmatively. The team agreed it would add updates on Lean Six Sigma to its weekly staff meetings and adjourned the meeting.

7

The Project Begins

"I'd like you to lead this effort" lingered in Mark Moore's mind as he drove home that afternoon. Based on the meeting earlier in the day, it was clear he had the leadership team's buy-in. He was particularly gratified to feel John Saylor's support as the CEO. Mark was excited to be leading this highly visible effort. The quality function always seemed to play second fiddle to the manufacturing and engineering functions; here was a chance to be in the driver's seat for awhile. Excited about the new mission, he grabbed the Six Sigma book and began to study it more closely.

Based on his reading, several things seemed clear:

- Project success would depend on following the DMAIC (define, measure, analyze, improve, and control) process and utilization of the appropriate statistical tools.
- Six Sigma should be applied to projects that generate significant "hard dollar" savings.
- Engaging a Six Sigma consulting firm would accelerate progress.
- Developing qualified and trained leaders in the organization would be critical to the project's success. *TryinHard Marine* would need some "Belts" and soon!

The next day, Mark, a member of the local Chamber of Commerce, Rotary, and several professional quality associations, tapped his networks to identify potential Six Sigma consultancies. In keeping with *THM*'s procurement policy of evaluating at least three vendors, Mark contacted three firms that would serve the purpose for the procurement requirement. The first, *Serv-Qual*, had all the right credentials and an impressive client list, but was mainly focused in the service sector with little experience in manufacturing. The second firm, *Accounture*, was a large firm with a variety of clients and some focus on LSS. Its service niche was more in the accounting function, which Mark was sure would please Sally, but would not impress Warren.

Still, Mark liked the larger firms because "bigger must be better." "No one ever got fired for hiring IBM," he thought and ultimately contacted *SixSigma-R-Us* (*SSRU*). Based on his research and professional network, *SSRU* was the top LSS consultancy that worked with large manufacturing businesses. Mark arranged for a representative to visit him at the office as soon as possible.

The *SSRU* representative, William Sellya, confirmed Mark's initial thinking regarding Six Sigma. William related several stories about successes *SSRU* had obtained working with *EG*, *Moneywell*, *4N*, and other Fortune 50 clients. Multivariable statistical analysis and designed experiments helped identify the root causes of intractable problems, which had stymied the clients' engineers for years. Discipline was key, as was extensive training to develop the internal capability needed for *THM* to achieve its business objectives. The *SSRU* representative emphasized their ultimate objective was to transfer not only the skills, but the ability to grow and maintain a Six Sigma culture to *TryinHard Marine*. William shared his belief that the standard ratios of one Black Belt per 150 employees and one Green Belt per 50 employees were reasonable. They had worked for other *SSRU* large clients.

Mark asked about the role of *THM*'s senior leadership team. He seemed taken aback when William recommended Mark and his colleagues be trained as Green Belts. Mark rejected the idea immediately and conclusively. After all, he and his peers were smart, savvy, and extremely busy. They had a business to run! Mark did see some advantages to getting the leadership team together. Obtaining a better understanding of the projects and the commitment needed from each of the leaders made sense. It also seemed like a good opportunity to build stronger alignment among the staff. Mark told William Sellya that he had no doubt his teammates would be more than fine with a half-day accelerated version of the material. William, not wanting to lose a major sale, acquiesced.

Several conversations later, the deal was done. In the first phase of the project, *SSRU* would train one class of Black Belts, three classes of Green Belts, and provide coaching support for the first year. The contract totaled $350,000, with first-year annualized, hard dollar, project savings projected to be between $600,000 and $750,000. Since Mark was anxious to get started, he convened the leadership team for approval to sign the contract. Though Warren Mast was again traveling on business, the others decided that time was of the essence and agreed to begin immediately.

Mark was elated. His influence on the leadership team was clearly on the rise.

8

The Training Begins

SixSigma-R-Us's Ben Black, the Master Black Belt, was excited. Leading his first full-scale client implementation with *SixSigma-R-Us*, Ben looked forward to meeting the executive team and getting started. His colleague, William Sellya, gave him the bad news that he only had a half day with the leadership team. Ben was confident he could convince this group to spend more time once he hooked them on Six Sigma.

By the end of the half-day meeting, Ben was less certain. He had emphasized the importance of a critical mass of Belts and shared some characteristics that *THM* should consider in selecting these important and critical people. Everyone nodded politely, especially at the admonition to "pick your best and brightest as Black Belts." Karen (executive vice president of manufacturing) thought to herself, "The best and brightest available *at the time.*"

Two conversations in the working session were noteworthy. The first involved project selection. Ben had provided a list of possible project criteria. There were several, but the two he emphasized the most were (1) importance to the customer and (2) hard dollar financial return. Brian Rudd (sales) immediately started ticking off the complaints his reps had been hearing—backorders, poor quality, and billing errors were mentioned. Everyone around the table (including Warren, who had returned from his travels) became defensive. "Yes, these are issues, but they weren't caused by *my* function" was the comment made by several around the table. Ben played the role of mediator, arguing that the issue was with the processes, not the people. The room quieted and they agreed to think more deeply about initial project selection.

The second confrontational conversation surfaced during the Black Belt selection discussion. Ben shared the requirements: one week of training per month for five months for a total of 200 hours of training per person, plus working on a project. The expectation he established with the staff was that *THM*'s Black Belts would be certified in six months or less, shortly after completing their projects. Everyone's internal calculators started clicking. The collective thoughts around the table seemed to be, *"Is this guy nuts and is Roswell missing an alien?"*

TryinHard Marine did not have a slew of people-savvy, process-savvy, credible team players who were quantitatively inclined. In fact, the very few that came to mind were in critical middle-management positions and could not be replaced. All eyes shifted to Warren Mast in engineering. "Your manufacturing engineers sound like a great fit," said Karen. "And who knows our processes better than Quality?" asked Sally (finance), glancing at Warren.

Controlling his anger, Warren responded, "That's not what you've been saying for the last six months when I've asked for more staff. I still have two open requisitions hung up in an HR black hole, and now you want me to give away six more of my people?" He emphatically said, "No way!"

Mark Moore found himself in an awkward position as all eyes turned toward him. He wasn't sure his staff was "best in class" when it came to being perceived as credible, team-based problem solvers. On the other hand, the Six Sigma implementation had been assigned to him and it wouldn't look good to not support his own initiative. "This is important to our business and the Quality Division will be a leader. I nominate my best and brightest—Fred, Kathleen, Jean, and Chris—for Black Belt training."

Now it was Karen's turn to get uncomfortable. She knew Mark's team and felt he was not being totally honest. Karen thought that while Kathleen clearly fit the criteria and would be an excellent Black Belt, Fred and Jean were competent at best and Chris was a problem. Karen had been planning to speak to Mark and Paula Lennox about his arrogant attitude, but hadn't had a chance to do so. Yet this meeting didn't seem like the time or place to raise a personnel issue, and Mark was showing leadership in providing candidates, so she kept her concerns to herself.

Ben Black turned to Karen next. "Karen, you control the largest number of employees in *TryinHard Marine*. Who do you nominate for Black Belt training?"

Karen responded, "I was thinking my organization would provide the Green Belts. I'm not sure I have people with the combination of skills you're looking for in a Black Belt candidate."

Ben smiled and said, "Sure you do. Consider this as a leadership development opportunity to develop your best people. Organizations often promote their Black Belts to middle and upper management positions due to their Six Sigma knowledge and experience."

Paula from HR then spoke. "We already have a leadership development program in place that works very well. We'd have to do some major adjusting to add Six Sigma into the mix." Karen wasn't at all sure Paula's program worked as well as she believed it did, but again, this didn't seem like the place to bring it up.

"Ben, I think I could find one person from each manufacturing location to become a Black Belt, but they can't be full time. I'll commit to the five weeks of training, but I need them to be able to perform their daily jobs as well as complete their projects. Can you live with that?"

Before Ben could respond, John Saylor broke in. "Excellent thought, Karen. We're committed to Six Sigma and we're also dedicated to running the business and satisfying our customers. It wouldn't be fair to them to take our best and brightest away from the jobs they do so well right now. These are bright, caring, motivated people. They will figure out a way to get both things done."

Ben Black was concerned, but not too much. Once *THM* got a taste of Six Sigma results, investing in more training would be a no-brainer. If starting with part-time Black Belts would get this account going, he could live with it.

At this point, Paula Lennox asked which line item would be charged for the training. Sally Dow, CFO, spoke up and snapped, "That's not your problem. I've committed to John and the team and I'll find a way to make the up-front investment work." Paula shrank back into her seat without a reply.

After several follow-up meetings, the leadership team eventually identified a class of 12 Black Belts. Five would come from manufacturing, four from quality, and three (with a great deal of grumbling by Warren Mast) from engineering. The initial wave of Green Belt candidates was also selected, with the majority coming from Karen's manufacturing organization as expected. Interestingly, neither HR nor the Controller's office volunteered people for either Black or Green Belt training. Each felt that their staff would only be used on an as-needed basis to answer specific questions so the time away from work to attend the training would not produce a positive ROI. Brian Rudd from sales also demurred. He said that, based on his research, sales and marketing projects did not lend themselves particularly well to Six Sigma approaches. He left the door open to participation in subsequent waves of training.

The leadership team also (eventually) selected the initial projects. Ben had asked them to identify a project in each facility so that the Six Sigma culture change could start simultaneously in all parts of the business. Karen resisted, concerned she couldn't afford to have every facility potentially disrupted by Six Sigma training, projects, and learning curves at the same time. Mark agreed, suggesting that a small number of projects at the home office be selected as "proof of concept pilots." In this way, the leadership team could keep direct tabs on what was going on and provide appropriate support. John Saylor nodded his head and asked his staff to focus on San Diego-based issues.

Three projects were selected. The first focused on excess scrap. Ben had recommended this type of project because of its narrow scope, easy measures, and hard savings.

A second project focused on excessive variation in vibration on one of the wiring machines. Ben had asked the group where there was excess and unwelcome variation from an important standard and the wiring machine was certainly a candidate. Warren Mast mentioned that he had a team working on the problem for awhile. The group jumped on this, stating that the team could use the data the engineers had already collected to accelerate the process improvement. Warren thought to himself that it would be better to just let them finish the project, but he said nothing and the team moved on to identifying the next project.

The third project focused on reducing customer complaints on the OWow123 NavSystem. Ben kept pushing for customer-focused projects. Given the increase in complaints and the introduction of a new competitive product by *GoGetem*, this project seemed like a no-brainer. Paula asked if the project was too broad, a comment swiftly dismissed by the others. Ben tried to make peace, commenting that many times the first job of a new team was to gather initial data and use it to narrow the scope. Paula nodded her head, grateful for any crumb of support.

After the meeting, Ben Black summarized the results for himself. The CEO had been supportive, but didn't seem ready to put Six Sigma near the top of his list of priorities. The CFO had asked some challenging questions about when they would be seeing savings, an expected and good thing. However, she balked at his suggestion that training some finance people as Black or Green Belts would be a huge professional development opportunity for them. Karen, the manufacturing leader didn't seem very comfortable with her Black Belt nominees and there were some interesting nonverbal reactions from others in the room. That would make his job more difficult. Finally, the vice president of engineering seemed resistant for a reason Ben couldn't fathom. "Well, those are the challenges that come with leading a major project," he thought.

The Black Belt (BB) training lived down to Ben's expectations. Out of the group of 12, there were probably four who really belonged there. Kathleen, as advertised, was a super-star. Two of the engineers seemed to have both the quantitative and people skills needed for success, and the manufacturing manager from the Miami facility had real potential. The others would be a real test of his instructor abilities.

One individual, Chris, was a particular challenge. He seemed determined to make the training a contest about who was the smartest. He never missed an opportunity to put down another classmate or to confront Ben about what he perceived to be an inconsistency in his presentation. He even pointed out typos in the printed materials. The rest of the group seemed to tolerate his outbursts. "That's just Chris," commented several participants during a break. While on another break, Ben asked Chris to meet with him at the end of the day.

Chris replied, "Sorry, I will not be able to meet today. I am too busy and must leave immediately after the workshop." Ben then asked him to meet prior to the beginning of the next day's class. Chris agreed.

That evening, Ben reflected on his upcoming conversation with Chris. He knew how important it was for the BB candidates to be taught all of the course materials. They would be expected to know the information for tests and as a requirement for certification. The classes needed to stay on track. Individual disruptions could not be tolerated. He then developed a strategy for use the next morning.

The next day Chris, true to his word, arrived 15 minutes early to meet with Ben. Ben opened up the conversation, "Chris, I am glad you found some time to meet with me. As you may or may not know, it is critical that BBs learn the course materials quickly. There is an aggressive timeline and schedule necessary so the teams have the information they need to begin working on their projects. Are you aware of this?"

Chris replied," Yes, but this stuff is fundamental to my training and education. I think some of us could just jump ahead and get to work on the project rather than go through this rudimentary retraining. It seems we may have jumped the gun in hiring outside consultants to teach materials we could have easily provided."

Ben, sensing competition with Chris and not sure where to go with this comment replied, "Well, I understand you might feel as if you could teach this class, but *THM* made the decision to bring in *SSRU* and here I am. How can we work together?"

Chris replied, "I am not sure. What did you have in mind?"

Ben, thinking this was his chance to get Chris to stop disrupting the class, said, "Well, to start, I would ask that you don't belittle the materials or others who are trying to learn. In turn, now that I know you have been exposed to these materials, I would appreciate your help in translating them as they apply at *TryinHard Marine*. I would like us to work together and not be at odds for the sake of the work. Can you agree to do this?"

Chris, still annoyed, begrudgingly agreed to work with Ben—for now.

The classes continued, with Chris toning down his sarcasm. As a result of this mix of participants, getting through the prescribed content each day became an ongoing challenge. *SSRU* had a time-tested training formula and Ben was expected to follow it: one lesson per day, five lessons per week. Ben knew from experience that if the group started falling behind, catching up would be difficult. So, he did what he needed to do: cajole, threaten, coach; whatever was needed to get through the day's materials. Yes, the materials were challenging; however, being a Black Belt was an honor. Black Belt certification provided a level of visibility unheard of for these participants and a chance to make a large monetary difference to *TryinHard Marine*. If they didn't want to rise to the challenge, they needed to leave.

9

The Training Drags On

"Why are we here and who did we make angry?" Those were the questions as the Black Belt (BB) candidates commiserated that evening after a long day with Instructor Ben Black Belt. The day had focused on the details of statistical sampling. There was a lot of material to cover and Ben had driven them hard to get through the assigned pages.

"What did we do wrong to be punished like this?" asked Rob, the manufacturing BB candidate from Seattle. "I was minding my own business, doing a good job, and all of a sudden I'm here being trained as a graduate-level statistician. I mean, I can balance my checkbook and read an operations report, but sampling, alpha and beta errors, and power? Please, we will NEVER, EVER use this stuff outside of class exercises. And why do we have to do the calculations manually? Oh, yes, because Ben thinks it is: (1) good for us to be hands-on with the concepts, and (2) what if the computer network is down? If our system is down, trust me, you can bet I won't be working on beta." The others cheered and Chris bought Rob another beer.

Chris then commented, "It does seem we have tried these big kickoffs in the past and they just don't work. Nobody cares after the big kickoff has come and gone, and then my department is held accountable for the poor implementation and application of the quality processes. We were never asked to help with this and the whole thing feels like an added burden."

The group began airing a round-robin of complaints regarding the Lean Six Sigma (LSS) training. There were several comments about Ben's narrow focus and the level of disconnect between the content and the real world. Each took his or her turn complaining. Kathleen found herself in a precarious situation. She really *did* think this was important, that customer-focused, data-driven, systematic process improvement was vital to

TryinHard Marine's long-term survival and prosperity. She realized the others weren't there as of yet and kept her personal views to herself while she added a small complaint into the mix in a show of solidarity.

Things were not faring much better in the Green Belt (GB) classes. Ben had the same mandate: to stay on track and get the participants through their prescribed pages, tools, and lessons. He observed the usual "bell curve" of GBs. There were a few superstars—several that were more skilled than many of the Black Belt candidates. Ben felt the high number of duds (people who were there to occupy seats to fill a departmental quota) was representative of how their managers really felt about Six Sigma and quality improvement.

The GB class was getting through the materials, and the interim test scores were consistent with a normal distribution. Ben had offered individual tutoring to those with the lowest scores. A couple of participants had taken advantage of it, and he was confident they would be okay. The ones who didn't respond to his offer? Well, you can lead a horse to water but you can't make it drink. In Ben's mind, the majority of participants were progressing normally. He was not overly concerned about the teams' ability to deliver the promised project results. Ben slept well that night.

10

The Projects:
An Update

The Excess Scrap Team was feeling pretty good about things. The project's definition and mission were defined, quality data had been collected, and the root cause analysis was thorough. The initial pilot of their proposed improvements showed real signs of making a positive difference and controls were being developed. Karen Decker had taken a sincere interest in the project and considered scrap a big deal. She viewed this as an opportunity to finally eliminate a troublesome problem. Ben Black responded positively to Karen's interest in the team and provided some very practical suggestions for scrap reduction. This is how Six Sigma deployment was supposed to work.

The Excessive Vibration Team was not faring as well. Warren Mast, champion for this project, seemed to take its very existence as a personal and professional insult to his department and his own individual skill as a leader. As a result, he was extremely critical and nitpicked every action taken to move forward. This was especially distressing because, over the course of Warren's 25+ years with the company, he was known as a mentor to groups and individuals. The vice president of engineering certainly had the technical knowledge to provide guidance and perspective. In addition, the project fell in his area of responsibility. The team's success would help him hit his own performance goals.

Despite Warren's occasional speeches to the team that the project was important and that he was dedicated to its success, his actions belied his words. He did not seem open to the possibilities of new options for solving the problem. Members complained that his leadership certainly did not feel supportive.

Further evidence was provided by the lack of engagement and participation by the two engineering BB candidates who participated on the team. Based on overheard comments, the other team members concluded that the engineers believed they already knew how to solve this problem. They were just waiting for the other members to hurry up with the analysis, which would validate their answer. Warren seemed to ignore the lack of cooperation by his engineers and always supported their positions and behavior.

Putting things in perspective, the Excessive Vibration Team was in great shape relative to the customer complaints project. Talk about confusion and lack of focus! The Green Belts felt they were still in the starting blocks. No one seemed to be able to decide what this project was supposed to accomplish. The definition of *customer* changed by the day, depending on which vice president happened to drop in. Every Black Belt wanted, or *needed,* to gather different data to support their hypothesis about what kinds of complaints they were supposed to be addressing.

Ben Black tried to help, but was powerless to bring about a consensus among the warring factions. On top of this, Chris, the problem BB candidate during the training, had morphed into Napoleon Bonaparte and was barking out orders with acid and abandon. The environment was so bad that two of the Green Belts began a half-serious conversation about leading a mutiny and refusing to participate in the remainder of the training. Almost three months after the project began, the team did not yet have a signed charter and was still struggling to move from the Define phase into the Measure phase of the project.

Ben realized a change was needed. He approached Mark Moore with a detailed list of his concerns about Chris. He provided specific data regarding Chris's disruptive behavior during the training as well as during the project. Ben used the analogy of a toxic waste dump which, if allowed to spread, could be fatal to the team. Mark was in a quandary, as Chris reported to him. Mark was the one who recommended Chris as a BB candidate (though in fairness, no one else had raised any concerns).

Mark told Ben that he would have a heart-to-heart conversation with Chris. With uncharacteristic assertiveness, Ben replied that the time for talk was long gone. Ben recounted multiple conversations he had had with Chris to no avail. The team desperately needed a "kick in the pants"—a dramatic, symbolic move to illustrate the need to change its way of working. After a long pause, Mark realized Ben was absolutely correct. The next day, Chris was excused from his Six Sigma responsibilities so that he could "work on another high priority project for *THM*."

Ben figured he now could at least point to one small success.

11

A Staff Meeting

Five months into *THM*'s Six Sigma implementation, John Saylor was not a happy man. It seemed that only a few of the expected benefits of Six Sigma were materializing. He remembered the original meeting where the leadership team decided to move forward and used the key points as an internal scorecard (Figure 11.1).

John called a leadership team meeting to get a sense of what his staff was thinking. It took less than five minutes for the finger pointing to begin in earnest. Each of his direct reports had plenty of reasons why *THM* was not making the expected amount of progress, and every single one of them explained why the difficulties were based in someone else's area.

The person who received the greatest amount of grief was Mark Moore, vice president of quality. After all, this initiative had been entrusted to him. John could see the struggle taking place in Mark as he endured the barrage of negative feedback. Clearly, Mark supported Six Sigma. He had proven it with the leadership he provided in offering his staff as Black Belt candidates. Given his busy schedule, Mark attempted to drop in on the various team meetings and worked hard to maintain control over the contract with *SixSigma-R-Us* (*SSRU*). And yet, there was no denying that somewhere along the line, something had gone dramatically wrong.

That something needed to be identified and fixed, and soon. Brian Rudd had done a good job—perhaps too good—in raising expectations in the minds of key customers that they were going to see positive changes in several aspects of *TryinHard Marine*'s processes and practices.

Paula Lennox from HR and Mark Moore had done an equally effective job of raising expectations within the workforce. Paula mentioned increased stress was starting to show up in the form of an increased number of union grievances. She reported she didn't think it was critical (yet),

THE PROMISE	THE REALITY SO FAR
It would result in higher quality and lower cost, allowing *THM* to compete with both *JapanRocks!* and the growing low-cost threat of *BetterThanYou*.	No sign of savings yet. The Excess Scrap Team was getting close to implementing their pilot; the concept looked very promising. The Excessive Vibration Team was just beginning the Analyze phase, and the Customer Complaints Team was effectively nowhere.
LSS offered sales with an immediate story to tell their existing accounts, providing a reason to ask for a meeting and get some face time.	And they did share the good news. Brian Rudd did a nice job of creating an entire minimarketing strategy based on implementation of Six Sigma. Recently, though, clients had been asking about *THM*'s progress, and John had been doing some "dancing" with his answers. The lack of Six Sigma results reinforced the customers' perceptions that *THM* "over-sold" and "under delivered."
While causing short-term stress in manufacturing, Karen would eventually be the beneficiary of better quality and more controlled processes. It certainly seemed like a better solution than supervisors jeopardizing relationships with valued employees by demanding greater work output.	It became evident that everyone underestimated the amount of stress created by the Six Sigma implementation. Many key metrics were trending the wrong way—quality and on-time delivery being critical. Karen was getting increasingly defensive and becoming cynical, not her usual can-do style.
The first wave of projects would begin generating a positive ROI in three to six months.	Sally, who had been a genuine and enthusiastic supporter at first, was getting increasingly agitated at the negative cash flow she was experiencing. *SixSigma-R-Us* was certainly prompt in presenting their bills, but no one was willing to make a firm commitment as to when the promised offsetting savings would begin to appear. Secretly, she wanted to stop authorizing payments or at the very least slow payments until the initiative showed some bottom-line results.

FIGURE 11.1
John Saylor's LSS plan versus actual chart.

THE PROMISE	THE REALITY SO FAR
By signing up with one of the larger consulting firms, they would have a partner who had already been through the LSS learning curve, so their implementation should be relatively smooth.	*SSRU* was certainly a key player in the Six Sigma consulting space. John second-guessed his decision and wondered if they might be too big for *THM*. They had their prescriptive agendas and they were sticking to them, regardless of the increasingly urgent stream of feedback that was starting to reach his office. Most of the Belts he engaged in casual conversation respected Ben Black. There were positive comments about his technical mastery of statistics and the methodology. However, his lack of options around the training process was beginning to create a lack of confidence in his ability to see *THM* as unique. A few commented his "glory days" client stories of other organizations were starting to get old. Bigger and more standardized did not necessarily mean better, he thought.
Bonus thoughts: In a recent CEO Roundtable, other CEO's had been starting to wonder if Six Sigma was a good strategic possibility for their business. If he were to implement LSS at *THM*, it would be a chance to gain some "bragging rights" and move past talking and wondering into action.	John certainly exceeded his bragging rights in the first couple of months, enjoying the good wishes of his peers. While he didn't want to admit to them he was struggling, he certainly did want to find out if others were having the same experience as *THM*. He'd have to figure out how to ask without sounding like an incompetent leader, or worse, a total loser.

FIGURE 11.1 (CONTINUED)
John Saylor's LSS plan versus actual chart.

but everyone would be well served to lower the stress levels in the organization for awhile. She also noted the increased complaints were only coming from the San Diego facility where the Six Sigma deployment was active. The other facilities, with the exception of complaining about some of the best people being gone too often for training, seemed to run as well as ever. Rumblings of a hostile work environment due to forced overtime and long hours were beginning to catch her attention. This was an organizational bullet she wanted the leadership to be proactive in dodging.

Warren Mast from engineering continued to perplex John. A conversation a couple of weeks ago indicated Warren was still angry that the team hadn't waited for him to be present before agreeing to embark on its Six Sigma journey. John empathized with him for awhile, then finally told him he needed to get over it and move forward. Warren agreed to try and the meeting adjourned. John was especially disappointed when he overheard a conversation where Warren had asked one of the Vibration Team members for some details that John could not fathom would be relevant if Warren were acting appropriately in his role as champion.

After the staff meeting, John began to second-guess his Six Sigma strategy. He silently questioned his leadership team's ability to work together at an executive level and lead their respective areas through this business change. Did he have the right people in place? Did they have the right skills? Right now, he wanted—no, needed—the results he had envisioned. John thought, "I should have ... ," or "Could I have ... ?"

"As usual, 20/20 hindsight is always perfect! How do we move forward?" he wondered. With this much time and money invested, he decided to "stay the course" for a little longer.

ONE YEAR LATER

John Saylor assessed his strategy for growing *THM* and was underwhelmed by the results. *JapanRocks!*, *THM*'s Asian competitor was making significant inroads into *THM*'s market share. Customer complaints were up. There was tremendous dissention in his own leadership team, so much so that the Board had commented on it. He was not surprised by the letter in his hand, the one announcing Mark Moore's intention to leave *THM* in eight weeks to pursue other career interests.

It appeared that the Six Sigma implementation fell short of expectations. John was disappointed with the results from the first group of projects. It felt like (even though he knew it wasn't true) nearly every problem could be traced back, either directly or indirectly, to the LSS initiative. For the umpteenth time he tried to figure out where *THM* had gone wrong. The business reasons to implement Six Sigma still remained sound. Who wouldn't want more controlled processes and less variation? How could anyone argue with developing the internal capabilities in the organization to better focus on customers and use data? As CEO, hadn't John himself asked his team to commit? Hadn't he cajoled, coached, been tough when he had to be, "stayed the course" during the expected resistance?

He continued his inventory of expectations versus actual accomplishments. In fairness to himself and *THM*, there were some bright spots.

- At this point in time, *THM* now possessed 10 trained Black Belts. Twelve candidates began BB training a year ago. One had dropped out before the training ended. Another candidate's spouse became very ill and he left the organization to care for her.
- Ben had promised that the BBs would be certified in six months. Since certification required, among other things, the *successful* completion of a project, only two of the BBs were currently certified. The others would hopefully finish their projects and be certified within the next six months—three times longer than anticipated.
- Certified or not, the remaining individuals represented new capability in process analysis and improvement. He looked forward to seeing how *THM* could leverage their newly developed skills.

Projects provided a mixed bag of results. The Excess Scrap Team had implemented corrective actions and seemed to have good controls in place. They made an impressive presentation to the leadership team. Team members showed how they followed the DMAIC (define-measure-analyze-improve-control) process, displayed lots of impressive-looking charts and graphs and seemed genuinely proud of what they accomplished. Their annualized savings projection was less than their target number by almost 25%, but what the heck, some savings were better than nothing. John and his staff gave the team a standing ovation at the end of the conversation, something he could not remember ever being done before at *THM*.

The Excessive Vibration Team was a different story. They had moved crisply through the Define phase, doing a nice job with the project charter, current-state process map, and voice-of-the-customer analysis. The beginnings of the Measure phase also appeared to be okay, at least in the beginning. The team presented what seemed to John a reasonable data collection plan.

Warren Mast, who had been distant and reasonably supportive to that point, turned from Dr. Jekyll into Mr. Hyde. He began showing up at every team meeting, injecting sarcastic remarks, and exerting excessive influence on the group. John questioned both the amount and nature of Warren's participation. As project champion, he expected him to provide direction, not do the work or to get that involved. Feedback from others on the team made it seem as though Warren backed the ideas of his engineers on the team without question, and squashed those of everyone else. John was concerned enough to ask Ben Black, the Master Black Belt, about the dynamic.

Ben responded in what sounded like Six Sigma coaching double-speak. "The team is working its way through some team issues right now. I'm monitoring the situation, reminding them of their meeting ground rules and making sure they continue to follow the DMAIC process." Ben also mentioned he had tried to schedule a coaching conversation with Warren, but that so far, they couldn't find a time that worked for both of their busy schedules. John, who knew Ben had coached many teams, decided to trust his judgment and did not intervene.

The team continued to progress through the Measure phase into the Analyze phase. They seemed to be reasonably on schedule—a good thing. Disquieting feedback continued to surface from team members. Apparently, the level of participation had steadily dropped. At first, the rumor was only the engineers were talking, mostly to each other and to Warren. Later, the nonengineering members of the team stopped attending the meetings altogether. John was alarmed by this development and asked Paula Lennox to follow up with those participants. She reported back that they seemed to have legitimate reasons for their absences. Paula then added she believed there were ways around those "legitimate reasons" if the team members really wanted to attend.

As the team reached the Improve phase, the corrective actions sounded eerily similar to those put forth by Warren Mast's engineers in a memo dated about 15 months ago. John paused. He was certainly pleased there was a viable solution to the vibration problem. But did *THM* really need to have invested all the time and money in a Six Sigma team if all they were going to do was come up with the same ho-hum answer developed over a year ago?

A dysfunctional Customer Complaints Team was yet a third story. After a year, the team had accomplished nothing, except to bicker constantly and create bad feelings among many people. John scratched his head. How difficult could it be to identify and solve customer complaints? After all, that's why customer service exists in the first place.

Yet each time the team picked either a class of complaints or a customer segment to focus on, someone threw a monkey wrench into the works: there wasn't enough data, the data wasn't usable, the scope was too big, the scope was too narrow, there was a big contract involved. The issues identified couldn't be solved by the team. At Ben Black's request, John himself had attended a team meeting to express his confidence in the team and to reinforce the importance of the project.

As Ben predicted, the team wanted John to make the decision. Ben had coached him to push the decision making back to the team so they would: (1) learn about themselves as a group, (2) learn decision-making tools, and (3) have ownership of their charter. He did as Ben suggested, answering their questions with questions of his own. They didn't seem too pleased, also to be expected, according to Ben. John was thinking about using a tool he heard about years ago by General Electric called Work-Out. Perhaps this rapid decision-making process might unlock this team and push the responsibility back to them.

Brian Rudd's sales and customer service people continued to share with customers the good news of Six Sigma; *THM* needed to get traction on a project that touched those customers directly. John decided to move out of "coach" mode and make an executive decision that would focus the team and get them moving forward.

Financially, the Six Sigma implementation was a nightmare. Only the Excess Scrap Team had implemented corrective actions that promised to generate real savings. The Excessive Vibration Team continued to struggle to get their new process stable (probably due to the resistance by all of the nonengineers), and couldn't give a confident prediction of when *THM* would see a positive ROI. The Customer Complaint Team was a complete waste of a year's training and time. *SixSigma-R-Us* continued to present its invoices like clockwork. There was no arguing Ben Black had spent the time onsite. There were always plausible responses to *THM*'s concerns about *SSRU*'s performance—or lack of it. Sally Dow, the CFO, became increasingly agitated as her profit and loss statements continued to trend the wrong way. She started delaying the payments to *SSRU* and John did not argue. It seemed one of the few things he could control this past year with *SSRU*.

Summarizing the last 12 months, John asked himself a question he often used as an internal benchmark for his decision making. *"Knowing what I know now, if I had the chance to do it again, would I have implemented*

Six Sigma at THM?" With real regret, he acknowledged the answer was "no."

THREE YEARS LATER

TryinHard Marine was struggling. In the past three years, their leadership position in the marketplace had eroded. The OWow123 navigation system had been rendered obsolete by their San Francisco competitor, *GoGetem*. It was especially frustrating to learn they used "Design for Six Sigma" as a development tool. *BetterThanYou* continued to drive margins lower with their competitive pricing. *JapanRocks!* was taking market share at an alarming rate.

Six Sigma had pretty much faded away into oblivion. There were pockets of successes. Kathleen had evolved into the superstar Black Belt everyone had anticipated. Somehow, she earned the trust of the other employees and she helped drive real change in those projects in which she was involved. Unfortunately, the other nine Black Belts (now down to six, three had left the organization) did not appear to be using their hard-earned skills. Shaking his head ruefully, John realized that Six Sigma had become the latest member of the "program of the month" club. Any mention of it produced head shaking and snide remarks. There were no corporately sponsored active projects underway and, with the exception of Kathleen's work, little evidence of DMAIC or its tools being used anywhere.

At a recent CEO Roundtable meeting, John had heard about another process improvement initiative called "Theory of Constraints." He wondered if this might be an approach that would work for *THM*. He would look into it next week.

Section II Summary

The *TryinHard Marine* (*THM*) Six Sigma case highlights many of the dynamics the authors have encountered when working with clients attempting to implement Six Sigma (or other business improvement processes).

We hope our shared experience extends to you, the reader. To deepen your learning, reflect on the following questions, either individually or with a group of colleagues:

- Does the *TryinHard* case story resonate with you? For what reasons?
- What was similar in the way *TryinHard* implemented Six Sigma to how it was done in your company? What was different?
- *THM*'s expectations were clearly not met in the story. Were they met in your own business improvement process implementation?
- *THM* did a poor job of creating innovative ways to implement and sustain Six Sigma. How does your organization compare?
- Was your business improvement effort effective in communicating across all boundaries? Why or why not?
- *THM*'s product and service quality was expected to improve after the implementation of Six Sigma. Did it? How about yours?
- Did *THM* achieve the ability to be flexible in changing conditions? How agile was your organization to changing conditions during its implementation of Six Sigma?

- How well was alignment and buy-in achieved by *THM*? To what extent was there "real" alignment in your own organization for Six Sigma?
- Based on the story, do you believe the *THM*'s implementation will be sustainable over an extended period of time? How about yours?

Hopefully, you are laughing right now. Maybe you are crying. Regardless of your industry, we hope you are saying something like, *"It's as if Brenda Fake and Larry Solow were here. They captured the essence of our experience."* If so, we accomplished our goal. There are linear paradigms embedded in traditional Six Sigma deployment processes that create systemic, dysfunctional patterns. These patterns cut across industry segments, geography, and organizational size. These are the dynamics we highlighted in the *TryinHard Marine* case.

In the chapters that follow, we focus on how to use new frameworks and tools to help business improvement implementations succeed. We will continue to focus on *TryinHard Marine*, providing possible solutions and options for action in addressing the challenges faced by a business improvement process implementation.

Some of these frameworks and tools will explain how to:

- Leverage connections to improve change efforts.
- Look at self-organization as an effective way to achieve change and build alignment.
- Recognize that business improvement success is a matter of "best fit" for your needs and situation.
- Amplify differences to achieve alignment and uncover new possibilities for action.
- Adapt to uncertain and unknown variables in your implementation.

Section III

So What?—Take 2

12

The TryinHard Marine *Case Story Retold*

The *TryinHard Marine* case story is a dramatization of our shared experiences with process improvement implementation. We're going to have a "do-over," integrating concepts and tools from Human Systems Dynamics (HSD) and Complex Adaptive Systems. The story resumes after the leadership team agreed to implement Six Sigma and went out afterwards to celebrate.

At the bar that evening, John Saylor bumped into a fellow member of his CEO Roundtable group. Tamara Hart was the CEO of a midsized healthcare provider. She had been a pediatrician and midlevel administrator before ascending through the marketing ranks to her current role. She exuded calm and credibility. Tamara tended not to say very much during the CEO Roundtable meetings, but when she did contribute, it was usually very valuable. John listened carefully to her comments; she was right far more often than wrong.

John approached Tamara, offered to buy her a drink (which she accepted), and excitedly told her of his leadership team's decision to initiate Six Sigma. He was surprised by the less-than-enthusiastic expression that flashed across her face. He was even more surprised by her verbal response. "That's quite a decision. I've had some experience with Six Sigma. If you like I'd be happy to share it with you, though not here at the bar. Perhaps we could carve out some time during a break at our next scheduled CEO Roundtable meeting." John quickly agreed.

After his exchange with Tamara, John felt a faint sense of unease. He'd been expecting a much more positive response from her. Was there something they were missing? John kept his concerns to himself, knowing he would be meeting Tamara again soon.

At the next Roundtable lunch, John and Tamara resumed their conversation about Six Sigma. John began by saying, "Tamara, let me get right to the point. Am I missing something? After we parted ways a couple of weeks ago I had a sinking feeling about *THM*'s decision to implement Six Sigma. There was something in your reaction that left me uneasy and looking forward to this conversation. Am I off base here?"

Tamara smiled, and answered the question with one of her own. "John, I'm curious. What, specifically, are the reasons you chose to implement Six Sigma? What, specifically, do you hope to gain?"

John thought for a moment and said, "Given our new competitive situation, we need to improve our productivity and innovation in the marketplace. We believe implementing Lean Six Sigma will give us an edge in the market and will be perceived positively by our customers."

Tamara sat back in her chair, listened, and sighed. "Let me tell you why I'm asking. We implemented Six Sigma at *Stars in Health* for what, in hindsight, were mostly the wrong motives, and we did it at mostly the wrong time. As you know, *BigBadBlue* is our major competitor and we thought we needed to 'keep up with the Joneses.' They had implemented Lean Six Sigma and were gaining market share on some of our key services in our own backyard. We conveniently forgot that *BigBadBlue* has 10 times more employees than we do, had extra cash due to a recent divestiture, and enjoys regulatory protection that allows them to operate with a very different set of conditions than *Stars in Health*. Any of these sound familiar?"

John grinned sheepishly. Without him saying a word, Tamara had identified one of their major justifications for moving forward. "And I thought we were unique." he replied.

Tamara laughed and said, "So did we. Let me re-ask the second half of my question. What, specifically, do you hope to gain?"

John shared his reasons: better efficiency, new conversations for his business development agents, and increased market share. Tamara took notes as John spoke. "They all sound like wonderful objectives to me. What have you already tried in order to achieve them? Have you implemented other process improvement initiatives, such as TQM or Quality Circles?"

This time it was John's turn to sit back in his chair and smile. "Yes, we've tried both. In fact, we invested a lot of time and money in each. Honestly, I was disappointed in the results. There was an initial burst of energy, but then it somehow just quietly slipped away until it seemed like a chore."

Tamara nodded her head and leaned forward. "I don't mean to sound like a naysayer here, but, John, I'm compelled to ask you, what makes you think Six Sigma will be any different? At *Stars of Health* we did TQM, Quality

Circles, data-based problem solving, and others I've already forgotten and had similar results. Each time we assumed that the reason it didn't work was that we had chosen the wrong tool or done a poor job of implementation. So, we changed tools or changed consultants, and tried it again. I appreciate your sharing your prior experience with those tools, John. Now it's my turn to say, 'I thought we were unique.'" They both laughed.

For the next several minutes, both were silent. Each reflected on their past experiences in attempting to make positive, sustainable change happen in their respective organizations; each realizing the amount of resources expended that hadn't provided the return on investment they desired. John broke the silence. "Tamara, you said you implemented Six Sigma at *Stars of Health*. How is it working for you? Am I making a mistake here?"

Tamara replied, "It's working well and you might be making a mistake—to provide concise answers to your questions. We chose to break the pattern we recognized in our earlier process improvement efforts and approached Six Sigma differently. It was, and continues to be, a lot of work. I do think it is paying off for us, both in the ways we expected and with some pleasant, unanticipated surprises. Whether *TryinHard Marine* is making a mistake is not for me to say. Rather than diving straight into tactical considerations, I would suggest you and your leadership team spend some time reflecting on the larger patterns at work within *TryinHard Marine*. Determine if you're satisfied with them. If not—and here's the hard part—are you willing to do the work of getting out of your own personal comfort zones to learn and try new things? The old adage, 'Insanity is doing what you've always done, exactly the way you've always done it, and expecting a dramatically different result' resonates with me. We did the hard work, and it has paid off for us. You and your leadership team have to work through that together."

The PA system announced the afternoon sessions of the CEO Roundtable were about to start. John sincerely thanked Tamara for her willingness to share her experience. "This is my way of 'paying it forward,' John. I was given a lot of support by consultants who specialized in an area of study derived from the sciences of complexity and chaos. The field is called Human Systems Dynamics and it created new possibilities for constructive action. They made a useful distinction between the linear quality of the Six Sigma tools and techniques and the complex nature of the individuals, teams, and organization that would be impacted by the changes.

You might want to check it out. In the meantime, I'd be happy to stay in touch with you, answer your questions, and share my experiences, even though our industries, business models, organizations, and patterns are different. So take what works, adapt what you need to, and throw away the rest." John quickly took Tamara up on her offer.

"Any last thoughts before we turn into CEOs again?" John asked.

"Two of them, actually," Tamara answered. "First, what do you *really* know about Six Sigma? I thought I understood the concept (actually I was too proud to admit I didn't). It's a big deal, so make sure you know what you're getting into. My second thought is really a suggestion. Be brutally honest with yourself and your leadership team about what is really going on at *TryinHard Marine*. At *Stars of Health,* we blamed everything and everyone else for the state of our business. We had more excuses than claims. When we really turned over the rocks to look at the dynamics of our complex organization, we found the problems—and the solutions— had an awful lot to do with us, the leaders. Good luck on your Six Sigma journey, John."

John went back and thought about *THM*'s TQM and Quality Circles experiences. He definitely did not want to repeat those debacles. He was committed to doing something different.

13

The Next Staff Meeting

At John Saylor's next staff meeting, the leadership team covered the tactical issues of the day. The usual litany of topics was covered: *THM*'s competitors' continued pressure on margins, sales' request for new products, manufacturing and engineering complaints about lack of resources, and quality's continued concerns regarding questionable decisions being made on the shop floor. It was "just another meeting" until John brought up the issue of Six Sigma.

"Consider the topics we just covered. Six Sigma could help us with each of them; in fact I recall our discussing these benefits in our last meeting. Has anyone done more research on this topic?" To no one's surprise, Mark Moore's hand shot up. He was the one charged with finding a consultant to help *THM* get Six Sigma started. "Anyone else?" John asked. No other hands were raised. "Too busy, too soon, didn't know where to start"—were the explanations.

"Well, I've taken some time to learn about it," said John, "and I think we may need to adapt our initial thinking." A general look of confusion permeated the room. This was not the CEO's usual approach. "I thought of our past experience with implementing TQM. Those initial conversations sounded scarily like the one we had last meeting. Everyone believed it was the "silver bullet" that would solve all of our problems. We brought in a TQM consultant who promised us a simple, smooth, and profitable implementation. Anyone here believe that happened? Raise your hand." All hands stayed right where they were and, interestingly, all eyes focused away from John. "Anyone here think we got our money's worth? Did we get our promised culture change? Did the effort sustain itself beyond its initial implementation? Hands?" Groans and sighs were audible; the staff began to squirm in their seats. This was definitely not the direction they expected the conversation to take. John continued. "Given our TQM experience, I've realized that implementing Six Sigma is more complex than I originally believed. We need to slow down, consider this more realistically and holistically, and determine what must be done differently to get a better outcome than we did with TQM."

It was John's turn to be astonished as he listened to the responses. Mark Moore: "I thought we made a decision. I've already got a lot of time and effort invested in this project."

Brian Rudd: "I need that Six Sigma story to tell my customers. We can get immediate benefit, even before we start implementing."

Karen from manufacturing: "One of my husband's colleagues was saying he'd heard really good things about Six Sigma. We need some good things to happen here."

Sally, the CFO: "I haven't been able to stop thinking about the juicy ROIs reported by successful adopters of Six Sigma. If we do this right, we can save some serious money."

Even Paula from HR, notorious for not saying much, chimed in. "You know I've been pushing for employee development. We can utilize Six Sigma as part of our staff and leadership training; this could be an element that might actually be welcomed by the business. Now there's a welcome change."

John paused for a moment to collect his thoughts and let everyone calm down a bit. "I'm very impressed by your enthusiasm. It bodes well for a successful Six Sigma implementation. Even with such high energy, which, by the way, mirrors the early conversations about implementing TQM, I'm concerned we will initiate what will be perceived as another 'program of the month.'" We can't afford that financially, emotionally, or competitively. We need Six Sigma to be sustainable over the long term and that will require significant changes by *us*—maybe even more than for the rest of *THM*.

"Before we move forward with Six Sigma, I ask each of you to reflect on some questions I have been wrestling with. What are you, individually and collectively, willing to do differently to support its implementation?" John stepped over to the flip chart and continued to outline his assignment (Figure 13.1). "I would like to review this information as a group. I am specifically requesting one slide from each of you that includes both your individual and our collective actions around quality and process improvement effort. In the left column, list the behaviors exhibited in the past. In the right column, write what you feel are the new actions and commitments we need to be successful in the future. I ask you to be

PAST ACTIONS (To support Business Improvement)	FUTURE ACTIONS (To support Six Sigma)
INDIVIDUALLY	
COLLECTIVELY	

FIGURE 13.1
John Saylor's past and future actions assignment.

brutally honest in your assessment of our individual and collective leadership actions (or lack of actions) in the past. And even more challenging, don't write anything in the "future actions" for which you are not willing to hold yourself and others accountable. I expect you to spend some time on this; it's too important to turn into a BS exercise. Be prepared to share those commitments, along with your questions, concerns, and challenges, at our next meeting. Any questions?"

Everyone sat quietly digesting the assignment now on the flip chart in front of them.

John looked around the room. Each member of his leadership team seemed a little quieter, a bit more reflective. There were no questions. Mission accomplished. He looked forward to the next meeting. He also looked forward to his next conversation with Tamara.

14

Choosing a Consultant

Mark Moore left the staff meeting with more questions than answers. John had clearly gotten his attention, as much by the change in his usual way of handling these kinds of business improvement initiatives as by the content of his message. The concept of learning from the Total Quality Management (TQM) effort hit close to home. Mark had been responsible for selecting the consultancy that worked with *TryinHard Marine* (*THM*) on TQM implementation. The fact that it hadn't gone as well as anticipated remained a sore spot for him, as Mark took a hit on his incentive compensation bonus that year.

So, the question for the day was what to do differently in selecting a Six Sigma consultant? For the life of him, Mark couldn't see what was wrong with hiring the No. 1 or 2 Six Sigma consulting firm in the country, yet that was exactly the approach he used when engaging the TQM firm.

After a couple of days of intense internal conversation, informal networking, and blinding headaches, Mark decided he needed a different perspective. He identified the perfect person to ask—John Saylor. He felt good about his choice for two reasons. First, John seemed to be thinking differently than usual. Perhaps his different perspective could shed some light on Mark's predicament. Second, having been burned once for a bad selection, at least he could say he had consulted with John.

John was pleasantly surprised when Mark entered his office and asked for his input on the Six Sigma consulting decision. In the past, Mark would have taken the ball and run with it, sure that as vice president of quality, no one else could possibly have an informed opinion on anything having to do with the subject. After Mark shared the results of his current approach to consultant selection (stuck and confused), the CEO leaned back in his chair and paused. He debated silently for a moment about whether to unveil his secret weapon—Tamara. John quickly concluded not doing so was the ultimate in petty gamesmanship, so he told Mark about the conversation he'd had with Tamara at the CEO Roundtable meeting. He recounted her distinction between the linear quality of the Six Sigma tools and techniques, and the complex nature of the individuals, teams, and organization that would be impacted by the changes.

Mark interrupted. "So did she find a consultant who 'got' both the technical and people aspects of Six Sigma and its deployment? If so, and she vouches for them, why reinvent the wheel?" John smiled, recalling he had asked a similar question during their lunch conversation.

"No, as a matter of fact, she didn't," John replied. "She ended up hiring two different consulting firms, one with expertise with the Six Sigma methodology itself and another specializing in Human Systems Dynamics, a field of study that focuses on organizations as complex adaptive systems."

"I wonder how that works," Mark remarked. "Two consulting firms with different mindsets, having to work together in a seamless and synergistic way. Sounds like a recipe for disaster."

John replied, "Apparently, Tamara made it the consultants' shared problem. She held them jointly accountable for the success of the implementation, making it clear that she didn't want to hear about any fingers being pointed at the other firm. Tamara told both parties that they needed to use their respective strengths and role model how these two dynamics could work together to make positive change happen at *Stars in Health*. While she didn't show me the contracts, apparently they included language that made it very clear that teamwork was expected and that there would be consequences—good or bad—for both firms depending on how successful they were—or weren't."

Mark sat back in his chair, a little dazed by what he had just heard. He had *never* heard of such an arrangement. Yet, if the contract was written the way John described, the combination seemed to offer the best of both disciplines. By making it the consultants' shared responsibility, Tamara had neatly removed herself from the "referee" role Mark often found himself playing when triangled between two warring subcontractors.

"Do you think Tamara would share the names of the two consulting firms she used?" asked Mark. "If they've already figured out how to collaborate for *Stars in Health*, we can take advantage of their learning curve. If they hate each other's guts, we know one combination of firms to avoid."

John smiled from ear to ear. "Way ahead of you, Mark. Here are the names of the consulting firms she used. She cautioned me that while they were an excellent fit for her healthcare issues and culture, there was no guarantee they would be a good fit for *THM*. In particular, her research found that different Six Sigma consulting firms seemed to gravitate toward certain industry niches. The HSD consultants appear to be less industry-specific. Tamara said it had something to do with their orientation toward patterns, which cuts across industry segments. I've already offered to buy her the lunch or dinner of her choice to learn more about this when our roundtable group meets again."

Mark took the business cards proffered by John with thanks. He laughed, "If this works out, I'll buy her dessert." Mark left John's office. He had some phone calls to make.

15

Establishing the Initial Projects

Mark Moore contacted the two consultants recommended by Tamara and after several conversations, engaged them both. Tina Grise, a senior consultant for *Successful Six Sigma* (*S3*), and Greg Sanchez, a consultant employed by *Human Systems, Inc.* (*HSI*) would be working with *TryinHard Marine*. John Saylor asked Mark to set up a conference with the consultants and to include him.

Prior to the meeting, John Saylor had called Tamara to ask for advice. Her suggestions were to listen closely to all parties, be mindful of his own biases, and to remember the reasons *THM* was undertaking Six Sigma in the first place. Based on Tamara's advice, John asked his staff to turn in their completed one-page assignment: (What will you/we do differently this time?) no later than the end of the week. After reviewing the information, John sent the information to both Tina from *S3* and Greg from *HSI* to review, consolidate, and present at his next staff meeting.

Two weeks later, the staff met to formally kick off the Six Sigma implementation process at *THM*. Tina Grice was introduced by John. She began by reminding everyone of the tremendous financial and other business benefits possible from a successful Six Sigma implementation. Tina discussed the need for a significant outlay of time, energy, and money. She did not downplay or skirt this issue, informing the staff that whatever level of resources they thought would be required should be doubled. "This is a real and major investment. I strongly recommend you tackle some projects that help you recoup some of it, recognizing you can't realistically recapture all of it at once." Sally Dow, CFO, nodded in agreement. This sounded much more realistic than the "positive return on investment (ROI) in six months" she'd heard about from others.

Greg Sanchez entered the conversation. "I agree with Tina. This *is* a tremendous investment of resources and *THM* certainly deserves to earn a positive ROI. I suspect that much of this is thinking you have already done as you considered the decision to implement Six Sigma. I am going to ask you to consider some additional thoughts as you begin your Six Sigma journey. These are derived from Human Systems Dynamics. In order for these additional criteria to make any sense, some background and definitions are essential."

Greg explained that it is important to differentiate between three kinds of change: static, dynamic, and dynamical. "Let me start with static. Think of a website that only provides stock, textual information to viewers. The text cannot change until the author makes a change. There is no interaction between the reader and the content. It is static. There are times static change is the right option. Updating information on a website is one of them.

"A second type of change is dynamic change, which involves a level of interaction. For example, when you toss a ball in the air, it will rise and fall based on the speed and height of the toss. The ball will change in predictable ways based on the variation of the two opposing forces of height and speed. This is dynamic. The opposing forces of supply and demand result in a dynamic market. Some popular organizational change models are dynamic in that they provide a clear step-by-step process to help people manage change. 'When you see this, do that.' But, what happens when the steps fail or a situation emerges not accounted for by the change model?

"Lastly, there is dynamical change. In this type, many variables interact and create self-organizing and emerging patterns. In dynamical change, many different variables interact in unpredictable ways. Individually, they seem random. Over time, though, a pattern often emerges, a pattern that could not have been predicted by any individual set of interactions. All three of these are present in complex adaptive systems, and each has implications for how to create positive change.

"I dropped a new term in the middle of that short lecture: complex adaptive systems. They are critical to understanding Human Systems Dynamics, or HSD, for short. Here is how the Human Systems Dynamics Institute defines HSD:

Human systems dynamics (HSD) is a collection of concepts and tools that help make sense of the patterns that emerge from chaos when people work and play together in groups, families, organizations, and communities.*

A special type of pattern is called Complex Adaptive Systems.

Complex Adaptive Systems (CAS) are dynamical, self-organizing, and continually changing. They are made up of "semiautonomous agents that interact in unpredictable ways such that they create system-wide patterns."† (Figure 15.1)

Greg continued, "Let me offer a graphic to illustrate this idea of semiautonomous agents. Consider that each of these circles is an agent, a player in the system. Those agents, each doing their own thing, create a larger pattern. That pattern (Figure 15.2), in turn, influences their individual actions. Visually, it looks like this:

These patterns are behaviors or events that repeat themselves over time and space. Patterns in time may be seen in cycles of economic growth, changes in customer requirements, and shifting sources of competition. Patterns in space may appear when teams in different places work together on a shared project or when information is accessible to part, but not all, of an organization. Different types of patterns influence organizations in different ways.‡

"Integrating these three ideas, HSD borrows from the fields of complexity, chaos, and other related theories to view people—individuals, teams, and organizations—as complex adaptive systems. These systems are nonlinear; they don't operate in the strict "cause and effect" way that machines do. Instead, they are complex, constantly evolving, moving in and out of what appears to be random, chaotic behavior. While we can't precisely control the outcomes of any given event—there are too many variables— we can recognize larger patterns that offer a way to understand and influence those systems."

---

* Human Systems Dynamics Institute Web site, http://www.hsdinstitute.org/about-hsd/what-is-hsd.html.
† Human Systems Dynamics Institute Web site, http://www.hsdinstitute.org/about-hsd/what-is-hsd/faq-tools-and-patterns-of-hsd.html.
‡ Human Systems Dynamics Institute Web site, http://www.hsdinstitute.org/about-hsd/what-is-hsd/faq-tools-and-patterns-of-hsd.html.

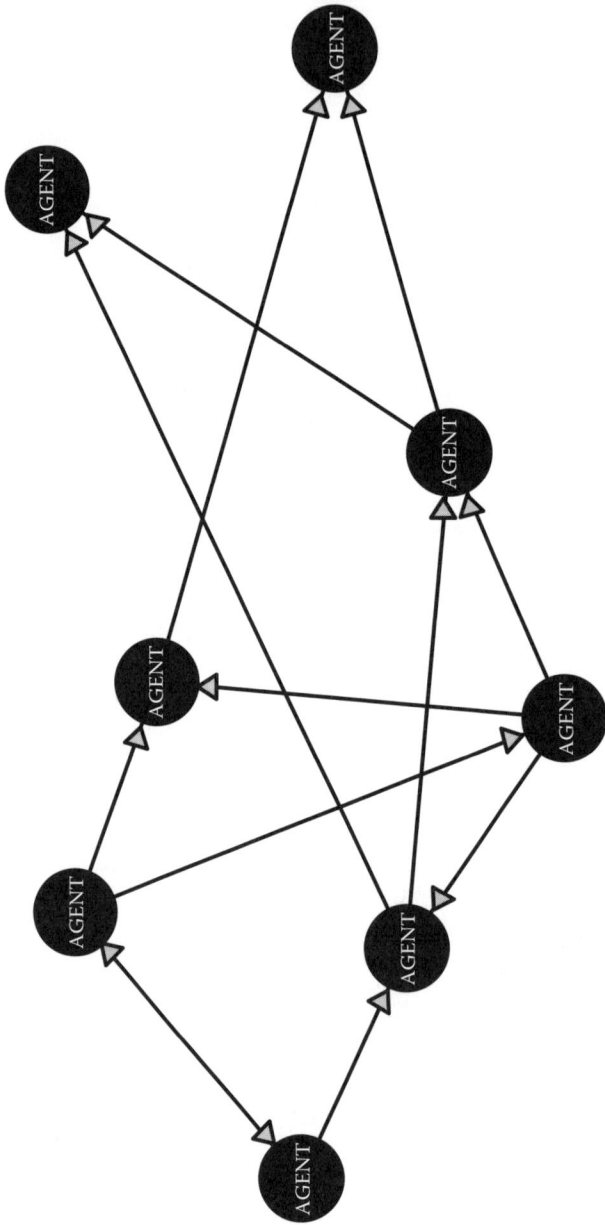

FIGURE 15.1

Semiautonomous agents. (Adapted from *HSD at Work: Frequently Asked Questions about Human Systems Dynamics*, Human Systems Dynamics Institute Web site, http://www.hsdinstitute.org/learn-more/library/faq-booklet.pdf. With permission.)

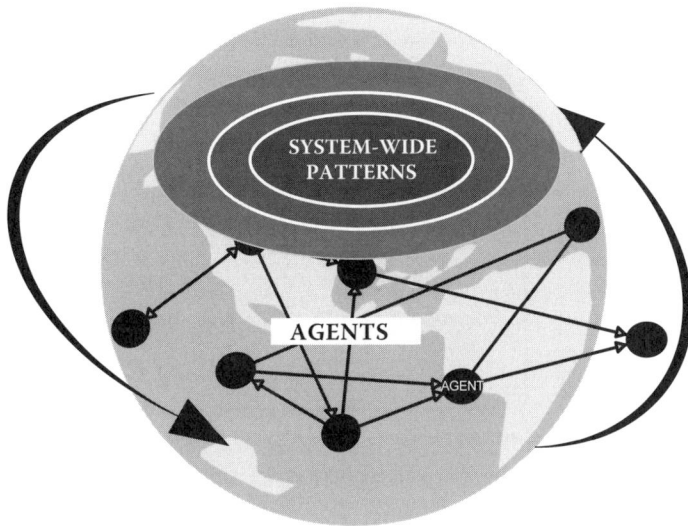

FIGURE 15.2
Complex Adaptive System (CAS). (Adapted from *HSD at Work: Frequently Asked Questions about Human Systems Dynamics*, Human Systems Dynamics Institute Web site, http://www.hsdinstitute.org/learn-more/library/faq-booklet.pdf. With permission.)

"The weather provides an example. While we can't predict next week's weather with certainty, we know that summer weather is generally hot, to be followed by cooler weather in the fall.

"You may be thinking, 'What does this have to do with implementing Six Sigma?'" Greg went to the whiteboard and drew a picture (Figure 15.3).

He described three different states a complex system could be in at any point in time. The "Organized" space is one where everyone is fairly certain of what the result would be of an action such that there is a high degree of agreement and certainty for an expected result. Everyone smiled when Greg offered an example close to the heart and wallet of everyone who worked at *THM*. "You all agreed to work at *THM* for money to be paid every two weeks based on your employment agreement. I'm guessing you all need to know with a high degree of certainty you can expect to see your paychecks deposited into your personal bank accounts every two weeks. This is an example of where an organization wants to have a high degree of control."

He described the "Unorganized" space next. The opposite of Organized, in this region there is little certainty about what the outcome of a given event will be, nor is there agreement about how to go about it.

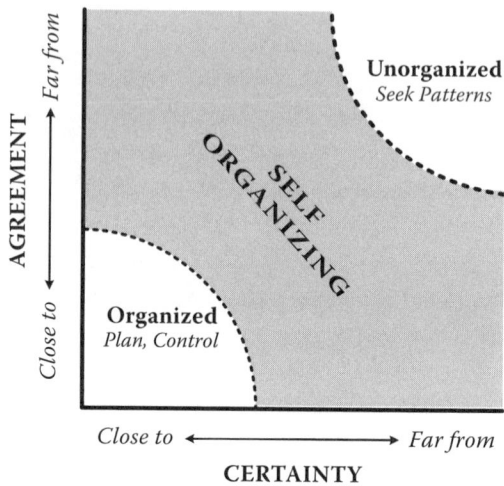

FIGURE 15.3

Ralph Stacy and Dr. Glenda Eoyang's landscape diagram. (Adapted from *HSD at Work: Frequently Asked Questions about Human Systems Dynamics*, Human Systems Dynamics Institute Web site, http://www.hsdinstitute.org/learn-more/library/faq-booklet.pdf. With permission.)

"Is a new product development initiative an example of this space?" asked Warren Mast of engineering. "We have a general direction, but no guarantee of success and as many suggested approaches as we have engineers; in fact, sometimes more because some people will come up with several." Everyone laughed, grateful for a moment of levity.

"Exactly right," Greg replied. Greg then asked, "Anyone want to hazard a guess as to what the "Self-Organizing" space is about?"

Paula from HR raised her hand. "I think it is the middle space between the two extremes. There is enough flexibility to allow the system to adapt to changes, but sufficient structure for things to get done. Our leadership development process comes to mind as a possible example. We have some broad structure and roles, but we recognize that each leader is different and customize within that structure to meet each person's unique needs. It's not like a free-for-all, though; there are certain concepts they must be exposed to or they won't be successful." Several members of the leadership team looked at each other with surprise. That was the most Paula had contributed to a meeting in quite some time—and she clearly got it.

After agreeing with Paula's example, Greg addressed the group again. "What do you think this might have to do with project selection? I'll give you a hint. Where are you now and where do your customers need or want you to be?"

After several moments of silence, Karen Decker, executive vice president of manufacturing, raised her hand. "If I'm following you, our customers want more certainty that we'll ship our orders when we say we will. They want us to provide them with increased confidence, increased *certainty*, that we'll honor our *agreement*, as defined by our promised ship date."

Brian Rudd from sales almost jumped out of his chair. "Shipping that way would save my sales and customer service people incredible amounts of time and heartache. They are constantly being put in the position of defending our inconsistent shipping performance. We're giving them ulcers."

Tina Grice reentered the conversation. "Moving processes toward the lower left-hand quadrant, the Organized space, is the place where the Six Sigma approach and tools are invaluable. The Six Sigma approach is all about identifying and reducing or eliminating variation. Eliminating that variation results in more consistency, greater certainty, and more *organization*. Warren, Design for Six Sigma (DFSS) also helps to move processes 'down and to the left' along the Landscape Diagram. In this case, it provides structure to the development of new processes to achieve 'robustness': the ability for a process to handle the widest possible range of parameters without needing to be modified."

Brian raised his hand, seemingly agitated. "But, what if we don't want to move 'down and left'? I need *more* creativity and customization from my sales and customer service reps. I think a lot of my people are stuck in a rut; stuck in 'the way they've always done it,' and those practices are no longer meeting *THM*'s or our customers' needs."

Greg responded, "Brian, you raise an excellent point, one that Tina and I would have brought up if you hadn't. Six Sigma is *not* the 'magic elixir' that cures all ills." To everyone's surprise, Tina reinforced the point, one of the first times anyone could ever remember one consultant agreeing with another that there could be a situation where their product was *not* the best answer. "Human Systems Dynamics offers a different set of tools to address ways to intentionally *increase* variation where needed—a conversation for another meeting," concluded Greg.

Greg smiled as everyone in the meeting cheered his last statement. "There is another HSD tool that is also relevant to project selection," said Greg. "Since projects don't—can't—exist in isolation (HSD calls this recognition of the 'whole, part, and greater whole'), it is useful to consider some distinctions that stimulate conversation about the connections between a given project and its larger context.

"HSD has identified seven components of 'dynamic fit' they've named the 7 Cs." Greg displayed a corresponding slide (Figure 15.4).

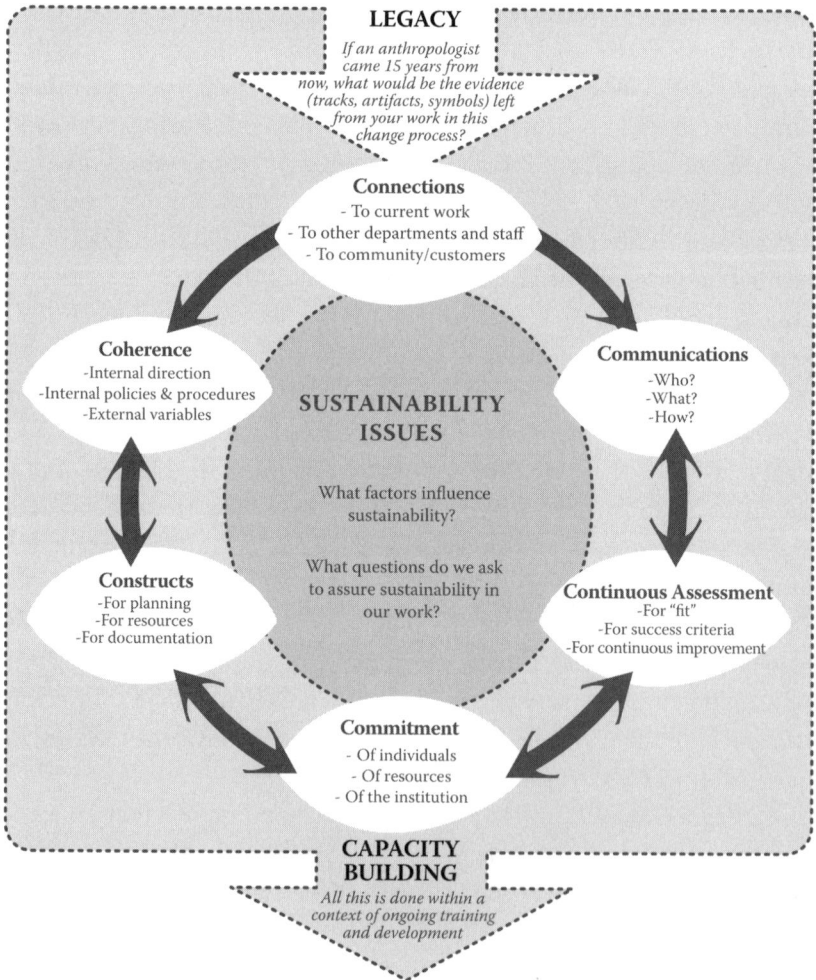

LEGACY

If an anthropologist came 15 years from now, what would be the evidence (tracks, artifacts, symbols) left from your work in this change process?

Connections
- To current work
- To other departments and staff
- To community/customers

Coherence
-Internal direction
-Internal policies & procedures
-External variables

Communications
-Who?
-What?
-How?

SUSTAINABILITY ISSUES

What factors influence sustainability?

What questions do we ask to assure sustainability in our work?

Constructs
-For planning
-For resources
-For documentation

Continuous Assessment
-For "fit"
-For success criteria
-For continuous improvement

Commitment
- Of individuals
- Of resources
- Of the institution

CAPACITY BUILDING
All this is done within a context of ongoing training and development

FIGURE 15.4

The 7Cs model. (Adapted from Royce Hollady, HSDP. *7 Cs Model.* Human Systems Dynamics. With permission.)

Each of the 7 Cs is a dimension of 'fit' in a complex system. They can be used in the context of project selection as follows:

1. If we were to select *this* project, what would its impact be on *Connections*? Do we have the supplier and customer connections available to us? Do we need to build them before starting?

2. If we were to select *this* project, what would its impact be on *Communications*? What information would we need to share with our various connections? Do we have reliable communication channels for information gathering and dissemination? What "story" do we need to share about the rationale for this project?

3. If we were to select *this* project, what would its impact be on *Continuous Assessment*? How will we measure the success of this project in both the short and long term? How would others measure our success?

4. How much *Commitment* is available to successfully complete *this* project? Are people at all levels of the organization willing to invest the resources needed? Whose incentive compensation will be impacted if this project fails? Will there still be commitment to "sustaining the gain" 12 to 18 months from now?

5. If we were to select *this* project, what would its impact be on *Constructs*? What tangible things—equipment, raw materials, finished goods, policies and procedures, ISO (International Organization for Standardization), and other controlled documentation—would have to be discarded, adapted, or created?

6. How much *Coherence* is there between *this* project and other projects and initiatives taking place in the organization? How well does this fit with organizational and departmental strategy? Other initiatives? Other projects? Is the sequencing right?

7. Finally, does *THM* have the *Capacity* to successfully complete *this* project? Are there enough people, with the right skills and time and other resources needed to do this project properly, using the full range of Six Sigma tools? Remember, not every project is a good candidate for Six Sigma, but for the ones that are, there is no better methodology. It *is* resource intensive. Are you prepared to invest?

The members of the leadership team had been taking notes furiously as Greg explained the 7Cs framework. A rich conversation ensued, co-facilitated by Tina and Greg, as the team grappled with selecting the initial set of projects that met the "down and left," 7Cs, and other, more traditional project selection criteria, such as time to implement and financial benefit.

Eventually they selected two. The first was reducing the variation between "promised" and actual ship date. The discrepancy cut across several departments, was important to the customers, and *THM* had already been working on the problem. Variation was clearly too high, so this project met the "down and left" requirement perfectly.

The second issue selected was more of a surprise. The project was to reduce the variation in accounts receivable for those clients whose terms were "net 30 days." Actual receipt of payments ranged from 30 to over 95 days, creating significant cash flow issues. At first, Sally rebelled, arguing that her financial people were "too busy keeping the financial ship afloat." After more conversation, though, the entire leadership team came to see this was a critical issue that directly or indirectly impacted everyone. Sally herself became a supporter as the magnitude of the potential savings (well over $1.5 million dollars per year) became obvious.

Tina took up the conversation again. "Now it's time to tackle the next part of our work for today. Having identified the initial projects, which individuals are you going to assign to them? Understand, we are talking about approximately 180 hours of classroom training plus an additional 40 to 75 hours of project team meetings, plus an additional 20 to 40 hours of individual work over the course of four to six months. It is a huge investment in the development of these people. Be sure you choose people who will provide you with a healthy return on that investment." Tina went on to talk about some criteria that had proven useful for other clients. She listed them on the whiteboard:

1. Comfortable with data
2. Good problem solvers
3. "Wide" and "deep" knowledge of the process being investigated
4. Perceived as credible by their peers
5. Have the potential to advance in the organization

Greg chimed in. "I agree with every single one of Tina's criteria. I'd like to offer some additional thoughts." Beneath Tina's points, Greg added:

6. "Systems" thinkers able to see patterns
7. Good networkers; people who "know who knows" and are comfortable collaborating with those gatekeepers
8. Comfortable with conflict; able to sit with differences without becoming (too) defensive or anxious
9. Comfortable with ambiguity; able to recognize that there are many possible ways to accomplish a goal
10. Flexible communicators; able to get their point across in many different ways; able to match their communication style to the person or group with whom they're interacting
11. Comfortable with "failure;" people who recognize that initial success is not guaranteed; people who have the ability to learn from what happened and try something different without getting (too) discouraged

Even John Saylor (who had been given a "sneak preview" of these criteria by Tamara during an earlier phone call) was sobered as he saw the list and heard the explanations. There were only a handful of people in all of *THM* who met most of those criteria, and all held critical positions within the company. Taking them out of daily operations for what was effectively 25% of their time for the next six months was asking a lot. Looking around, it was clear to him that his staff were having the same thoughts. "Here we go," John thought to himself. "Now we'll see how really serious we really are about this." And he waited ... and waited ... and waited some more.

The silence in the room was deafening. Tina and Greg waited as well, interjecting only to ask if anyone had questions about the criteria or had other inputs to add to the list (the latter statement received a nervous, "Are you serious?" response).

John caught Tina's eye and stood up to speak. "Remember, I asked each of you to consider what we as leaders, individually and collectively, need to do differently to get a better outcome with Six Sigma than we did with TQM. So far today, we have taken steps to look at the specific projects differently by using different criteria and including a finance project. That is new thinking, and I congratulate you for your courage. Now we are grappling with the challenge of who to select to work on these projects, knowing clearly it will impact current work assignments. If we are to really change our approach here, I ask again, what do we leaders need to do differently? I have asked Tina and Greg to look at your responses to the assignment I gave out at the last meeting (Figure 15.5). I would like them to offer their perspectives regarding what we can expect to do differently to ensure the success of this effort."

Tina showed the summary slide to the group. She and Greg had read each individual's responses, and synthesized them into themes. Greg noted that there was a high degree of consistency between the answers, and he felt confident of the summary they were viewing.

PAST ACTIONS (To support Business Improvement)	FUTURE ACTION (To support Six Sigma)
Individually	
Encourage my people to use the tools (Impatiently) ask what progress is being made Tell team members to call me if they need help or resources Celebrated project completion – showed up with "tokens of recognition" in appreciation for their hard work.	Make sure teams are given the time to meet. Show up from time to time at meetings where my staff are involved Make team updates a regular part of weekly meeting agendas Learn the tools (at least to some degree) myself so I can do a better job of coaching and reinforcing.
Collectively	
Agreed to spend the money on an excellent (high priced, anyway) consultant to support implementation Intervened when cross-functional issues arose Made sure TQM was highlighted in our monthly newsletter Projects selected based on specific team needs and desire	Make sure the teams are given the resources (time, maintenance, engineering, or whatever) they need to be successful "Get ahead of the curve," be thinking beyond the initial set of projects to what happens next. Do a better job of project selection. Select projects that support the overall business goals for the customer and bottom line

FIGURE 15.5
Summary of John Saylor's past and future actions assignment.

SUMMARY OF THEMES

Greg had written two questions on the flip chart:

1. What is similar between the two lists?
2. What is different between the two lists?

Responding to the first question, the staff members were able to quickly identify some common themes, including assuring sufficient resources, taking a personal interest, and solving disputes.

The conversation around the second question took longer. Karen raised her hand and noted, "It seems like our past actions had a 'once and done' feeling to them, where our future actions have more of a continuing, ongoing nature."

Brian cleared his throat and offered, "Adding to that, our past actions seemed more reactive, where our future actions are more proactive."

Tina commented, "Remember our definition of a pattern: a set of behaviors or events that repeat over time and space? Based on this conversation, how would you describe your 'old' pattern of supporting business improvement? Would the phrases, *reactive, once and done, not really investing of yourselves* be accurate?" Heads nodded yes. It was a harsh observation, but Tina had only paraphrased their words back to them.

Greg spoke. "What are the characteristics of a new, more 'fit' pattern to support Six Sigma in the future? Let's start by amplifying the differences and take the opposite of the old one." Greg wrote the following phrases on a fresh piece of paper:

- Proactive
- Continuing and ongoing involvement
- Making the investment of our own time so we can be knowledgeable and supportive

Greg continued, "Based on one of the comments, there are two more I'd like to propose." He added another bullet:

- Make the necessary investments of time and people
- Link process improvement to the rest of the business

Everyone agreed that both were also essential ingredients of a new, more effective pattern.

Tina joined Greg in facing the group. "One of the toughest decisions you have to make as a leadership team is how you deploy and develop your scarce and limited talent. Your old pattern has resulted in one set of decisions. We are challenging you to embrace a different pattern as you decide who will participate in your Six Sigma teams."

Paula Lennox from HR spoke up. "We've been talking about the need for a new model of leadership development. It strikes me that the ideal Six Sigma candidate criteria on the whiteboard are many of the same criteria we would look for in the future leaders of *THM*. We were prepared to spend big bucks to send them off to a fancy executive MBA program. Maybe we should consider these initial Six Sigma participants as our leadership development candidates. They will be exposed to data and people, linear process thinking and complexity. This group will need to work as a team, be individually accountable, manage their regular duties and their projects, and will be very visible to us. To the extent we choose the same caliber of staff we would send to an executive leadership program, the entire organization—us first—will know we are serious about Six Sigma."

John Saylor was pleasantly astonished. Two great contributions from Paula in one meeting. He thought this was the time to reinforce her efforts and thinking. "I think Paula nailed it," John said aloud to the group. "We can't afford *not* be dead serious about changing the culture of our organization. We spoke in an earlier meeting of our TQM implementation experience; not choosing the right people was clearly a contributor to our not obtaining maximum benefit from our investment. It's not like we're losing these people; they'll be right here, working on issues we just finished agreeing are critical to our customers and *THM*. As a bonus, since all of us will share the burden of investing in their development, we'll all be motivated to be sure they continue to use what they've learned after these initial projects are over."

John paused and silently looked each of his staff members in the eye. "Does anyone disagree with me? If so, I ask you to speak up or forever hold your peace." No one did. "Paula, I'm going to ask you to partner with Mark Moore in nominating—not choosing—possible team members. I request you run your potential candidates past their respective vice presidents for validation. If there is a legitimate concern, I need you to be open and realistic in acknowledging it, and then find another equally suitable candidate." John continued, "I'd like to see the proposed slate of team members before anyone is told about anything. Since, in effect, you're helping identify *THM*'s future leaders, I want to make sure I know who they are. Any reason you can't have that list of names ready for me by this time next week?" Mark and Paula exchanged quick glances and replied in unison, "No, sir. We'll have them for you." Their simultaneous answer evoked laughter from the rest of the team.

John spoke again. "Before we move off this topic, based on our TQM experience, here are some things I would like **not** to happen with our Six Sigma effort:

- The candidates should not be referred to as our "high potential employees." This sets them up as separate from the other employees and diminishes their ability to collaborate effectively, drive the projects, and make the necessary changes. The message must be clear and consistent; we are investing in them to develop skills. Others will be considered if these projects are successful.
- We are not going to start these projects with a lot of hype and fanfare. This effort is a part of daily work and improving our business performance.
- We cannot use Six Sigma as an excuse for not meeting our numbers. I do, however, expect everyone to communicate sooner than later if there is conflict between our projects and meeting our weekly and monthly commitments.

John offered some final thoughts to close the session. "This meeting was an excellent step in the right direction. I thank everyone for keeping it grounded and knowing we all will need to keep the higher good of the business in mind if this is going to work.

"For the record, I recognize the need to change my own patterns as well. If there is anything I need to do differently to help support this effort and your work, I expect each of you to communicate that to me directly. My bark is worse than my bite. Don't back down if you believe I haven't provided a sound rationale for my responses.

"Let's go forth and do great things and make some money." Everyone cheered and the meeting adjourned.

16

The Training

Paula Lennox (HR) and Mark Moore (vice president/quality) reviewed their prospective lists of the most qualified managers based on the criteria created by the staff. John Saylor (based on a hint from *Stars in Health* CEO Tamara Hart), also requested they identify the next most qualified at other levels of *TryinHard Marine* (*THM*). Tamara had told him the reason for doing this was to create a pipeline of Lean Six Sigma (LSS) leaders. Once the first wave of Black Belts (BB) was trained, the next waves would be better prepared for the formal training after being exposed to some additional duties.

Mathematical ability of the candidates was assessed, but only to make a rough assessment about the potential leaders' capacity to learn the needed statistics. Interestingly, Paula and Mark found that by defining two levels of qualified candidates, opportunities emerged for the next most qualified to grow by backfilling the Six Sigma team members. When Mark and Paula made their proposal to the others in the leadership team—complete with backfills—there were no critical concerns and the first class of Black Belt candidates was established. No one on the staff could remember a time when so many high-level people were selected with so little conflict.

The selected team leaders and members were divided into two project teams, one to work on the promised versus actual ship date issue (the Ship Date Team) and the other to work on accounts receivable variation reduction (the ARV Team). Both groups would attend formal classroom training sessions together so they could learn with and from each other as well as the instructors. Then, each team would meet separately with a consultant coach to apply what they learned in the classroom on their particular project. Consultants Tina Grice, of *Successful Six Sigma*, and Greg Sanchez, from *Human Systems, Inc.*, would co-lead the classroom sessions and rotate between the two teams during their working meetings.

As the training began, the BB candidates were struck by how process improvement was influenced by Six Sigma, HSD, or a combination of both. The participants learned how different kinds of customers, data, and approaches are needed when focusing on the linear versus nonlinear aspects of change.

"If you recognize that there are elements of both linearity and nonlinearity in your projects, you'll be better able to apply the right tools for the job," explained Tina. "Greg and I discussed—well, argued, if you want to know the truth—for quite awhile before we reached this understanding. I kept looking at the world through my linear, 'follow-the-steps' lens. Greg kept seeing everything as self-organizing. I'd tell him that a piece of testing equipment does not seek creativity and he'd respond that the operator of testing equipment can't be programmed like the machine."

Greg chimed in, "After a discussion over coffee that evening, we realized both of us were correct. Our job is to provide you with as many options for action as possible so you have a wide variety of choices available when confronted by the situations you will most likely face in doing business process improvement work."

And so the training progressed. They learned the linear tools of traditional Six Sigma, understanding how having and using data provided a "new lens" for dealing with problems that defied solution using their current problem-solving approach. In other sessions, they learned to appreciate the difference between a complicated linear system, with its many rules, replaceable parts, and predictability, and a complex nonlinear system characterized by simple rules, self-organization, and patterns.

One juxtaposition of linear and nonlinear tools was particularly striking. Tina first discussed the concepts underlying Design of Experiments (DoE). She explained how a designed experiment was a powerful way to test the results of multiple variables (she called them Xs) simultaneously on an output, which she named Y. Tina went on to explain the differences between main effects (the difference in outcome attributed to the change in any one variable) and interactive effects (the effect of multiple variables working together). She explained it using a comical video of the interaction between a carbonated cola drink and mints. "Pouring cola into an empty plastic soda bottle doesn't do very much. Dropping a mint into an empty soda bottle doesn't do very much, either. Put the two of them into an empty soda bottle and watch the jets of soda fly." She showed the video and everyone could see the dancing jets of soda spraying high into the air.

Greg stepped to the front of the class. "As we've discussed, individual outputs of complex adaptive systems by definition can't be predicted with certainty. They do, however, settle into patterns over time." He showed a video from Cornell's Professor Steven Strogatz,* which illustrated how a particular nonlinear equation settled into the "butterfly attractor." Regardless of a starting point, the results of the equation settled into what looked like two butterfly wings, even though it was impossible to tell where exactly the next result would occur.

Greg and Tina invited the class to compare and contrast the concepts underlying DoE and strange attractors. Participants shared examples from their work and personal experiences that fell into each group. Greg and Tina reinforced the point that the two have much in common. Both concepts recognize that outcomes are not always predicted or influenced by changing only one thing (although they might be). Both agreed that "the whole is not always the same as the sum of its parts," that combinations of variables often behave differently together. DoE and complex systems both use data, albeit different kinds of data, to help explain the patterns being seen. The two concepts recognize the importance of initial conditions (understanding what can and can't be controlled) and that these differences can make a significant difference in the results.

Greg moved to a different spot in the room and sat down in one of the chairs. All eyes turned toward him because he and Tina almost always taught at the front of the class in close proximity to the PowerPoint slides and their supporting materials. "What just happened?" asked Tina from her spot up front. "What is the same? What is different? How are you feeling right now?"

After a pause, Kathleen, one of the most outgoing participants, raised her hand. "What's different is Greg moved to the back of the room and sat down and he didn't say anything. What's the same is you're still up front, talking and asking us difficult questions." This produced some laughter in the classroom.

* "Chaos." The Great Courses series, #1333. The Teaching Company, 2008. Professor Steven Strogatz, Cornell University, Ithaca, NY.

"We've just illustrated a powerful tool for business process improvement," said Greg, speaking while still sitting in his chair. "It is a concept in Human Systems Dynamics (HSD) we've named *Same and Different*. One of the ways to influence patterns, to foster change, is to do what we call *amp* or *damp* differences. Amplifying differences magnifies them to create tension in systems for change (think pulling on a rubber band). Damping or minimizing differences (think about reducing the tension on an already stretched rubber band) reduces tension and allows the system to move toward greater stability."

Tina chimed in from her still-standing position in the front of the room. "Analysis of Variance tries to do a similar thing. It quantifies the impact of various elements of a given process, calculating weights to the element levels to establish which are statistically significant, meaning, which outcomes are statistically very unlikely to have happened by chance."

As the training continued, the participants continued to be exposed to both the linear and nonlinear aspects of business process improvement, learning concepts, and tools applicable to both.

Unfortunately, the same could not be said for the *THM*'s leadership team.

17

Another Leadership Team Meeting

About eight weeks after the training began, another leadership team meeting was held. Once the routine agenda items were concluded, John Saylor asked for an update on the Six Sigma deployment.

Mark Moore shuffled some papers and prepared to begin his update. He had expected this question and had some remarks all ready to share. He didn't get the chance. "I think we were sold a bill of goods," exclaimed Sally Dow, the CFO. "I know we are supposed to be patient, but I don't see anything happening at all, except the timely presentation of invoices by the *two* consultants."

Brian Rudd, vice president of sales, spoke next, though not with the same energy as Sally. "I *am* being patient and I'm hearing good things about the training from Kevin, my BB candidate. We're telling customers about deploying Six Sigma and *they* are asking when we'll see results. Not in a mean way, necessarily, but they are very clear they would appreciate improved quality. We were successful in raising their expectations, now we have to deliver."

Karen, executive vice president of manufacturing, raised her hand next. "My people are also giving high marks to the training. But, and this is a big *but,* they are concerned our productivity is taking a hit while trainees are in the classes. The team members appreciate that we've backfilled their spots, but they are spending a lot of time teaching, answering questions, and coaching their replacements. I've heard some mild rumblings that we didn't think this through well enough; that the backfill personnel should have been trained before beginning the projects."

Warren Mast, vice president of engineering, raised his hand and offered his assessment. "My engineers don't seem to be giving the same high marks to the training as the other BB candidates. That doesn't totally surprise me. For the most part, they already know the quantitative tools being taught to the others—probability and statistics are required courses in engineering curricula. They are also less enamored of the nonlinear Human Systems Dynamics (HSD) concepts and tools being introduced by *HSI*. Again, not a total surprise. In my experience, engineers generally self-select to work on neat, orderly processes. or to turn those that are messy into something organized as quickly as possible. They see the idea of introducing *dis*order into processes as sacrilegious, against the natural order of good engineering."

Both Paula Lennox and Mark Moore had been getting visibly agitated while listening to the litany of complaints. Sensing an explosion was about to erupt, John injected himself into the flow. "Before this conversation goes too much farther, I'd like to suggest we invite Tina and Greg (the consultants) to join us. It is important the consultants hear your concerns first hand. They may have already encountered and addressed these with other clients." Everyone agreed and John thought to himself, 'I owe Tamara a big favor. She told me this might happen and suggested I have the consultants available if needed.'"

After Tina and Greg joined the meeting, John had each of the leaders restate their earlier concerns. Paula raised her hand to offer a rebuttal and much to her amazement, Greg asked that she and Mark defer their comments for awhile and continued to write down the key points as they surfaced during the conversation. Greg stated, "What I think I'm hearing is that in some way or other, the implementation isn't going as smoothly as you had hoped; things aren't going 'according to plan.' Is that fair?" Heads nodded. "Do you think the concerns you've raised could have been prevented with better planning? Were you 'bad' planners? Shouldn't you have anticipated these issues and addressed them ahead of time? After all, you had the experience of the earlier TQM implementation. Maybe you're not as smart as you think you are."

The group was stunned into silence. This consultant was calling them out as poor planners, unwilling or unable to learn from their past experiences. Several leaders' blood pressure began to rise. Greg paused for a moment to make eye contact with each leader and then asked everyone to pull out a blank piece of paper and draw a line vertically down the middle. "On the right side of the paper, I'd like you to write down the key words I just said to you, as accurately as you can recall them. On the left side, please write down your reactions and feelings to what you just heard. This

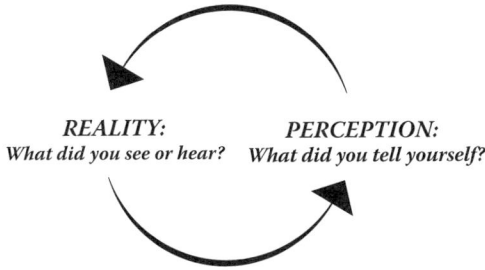

FIGURE 17.1
Tina's model of reality versus perception.

is not a group exercise; please do this individually and turn your paper over when you're finished."

When the last response was turned upside down, Tina stepped in to lead the debrief. "I watched Greg do this with another client and was very impressed with how quickly it helps identify the way each of us processes conflict—the pattern, the stories we tell ourselves when we identify a disagreement or problem as a conflict." She drew a model (Figure 17.1) on the whiteboard.

"Notice how our interpretation of the data—our perception of it—influences the reality we pay attention to in subsequent cycles. We 'see what we expect to see' is one way this is often described. A self-fulfilling prophecy is another." Greg continued to reinforce the point that organizations are complex adaptive systems, influenced by so many variables that it is impossible to control them all. "As a result, it is impossible to have a plan that goes exactly according to plan, unless that happens by luck. The process improvement skill (the leadership skill) is *not* in beating ourselves—or others—up when we aren't perfect planners. With complex adaptive systems, we can't be. Rather, the talent is to understand our internal patterns in how we deal with these setbacks and recognizing we have choices in how to address them."

Paula spoke up. "Are you suggesting that adaptability and agility are leadership competencies? We don't have these on our current list."

Greg said, "Right on, Paula; that's exactly what I'm suggesting. How could it be otherwise? Let me be very clear, though. This is not to say leaders shouldn't plan. In fact, planning is a vital part of your job. And in linear processes, or in the 'organized' space of complex adaptive systems, you should have a reasonably high degree of confidence your plans will play out as designed. The trick is to know which is which."

Warren Mast literally slammed his fist on the table. "No!" he shouted, amazing even himself at the passion he was exhibiting. He paused and took a breath. "Sorry, I didn't mean to get so carried away, but what you're saying runs counter to everything I've been taught in eight years of undergraduate and graduate engineering. You teach Designed Experiments to our Six Sigma candidates. With enough time and ingenuity, we can identify the variables that significantly impact the results we're trying to achieve. In effect, you're saying that can't be done!"

Greg kept his cool, watching Warren, listening hard, and paying genuine attention. "Warren, I can't tell you how much I appreciate and respect you for sharing both your views and your obvious commitment to the scientific method.

"Let me start by laying the foundation for a tool I will introduce in a moment to respond to your comment. While you may not have thought about this, it turns out there are three different types of truth. The first is factual. For example, the sun will rise and set tomorrow. This is verifiable and can be objectively validated. The second is subjective truth, a truth I know to be true, but you may not. Lastly, normative truth is present when several people share the same subjective truth. Keeping these in mind and, with your permission, I'd like to share another HSD tool that might shed some light not just on this particular issue, but on a larger pattern of conflict, this time a pattern that occurs between people" (Figure 17.2). Greg drew a diagram on the whiteboard.

We call this a *Decision Map*. Its premise is that when two or more parties see things differently, that is, perceive themselves as being in

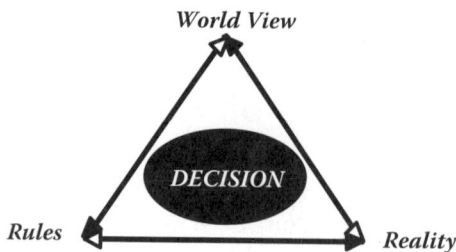

FIGURE 17.2

The Decision Map presented by Greg. (From Human Systems Dynamics, http://www.hsdinstitute.org/learnmore/library/faq-booklet.pdf. With permission.)

conflict, there is usually a difference in perception in one or more of these three areas:

- **World View:** We have different philosophies about what is important. For example, you were very articulate in your view of the importance of DoE and scientific method. This is the subjective truth I described earlier.
- **Reality:** Parties often differ in what data they perceive and how they weigh its relative importance. You noted, correctly, that we teach DoE to our BB candidates. You chose *not* to focus on the fact that we also taught the class about strange attractors. Both of these statements are factually true. And finally …
- **Rules:** There is often disagreement about how to proceed in a particular situation. In this situation, it seemed that your rule for addressing our differences was to use a powerful signal to get attention, then to articulate your position with passion. Rules, when followed by all the relevant parties, are examples of normative truth.

"The Decision Map describes a general pattern that we pay attention to when interacting with others in a conflict situation. Let me show this to everyone graphically" (Figure 17.3). Greg drew pictures on the whiteboard.

"We each have a map. In this case, two different ways we go about making decisions to resolve the conflict. Since in human systems there is a great deal of interconnectedness, changing one of the three elements of the Decision Map intentionally is very likely to have an impact on the others, even if we're not sure how." Greg paused to see if not only Warren, but the

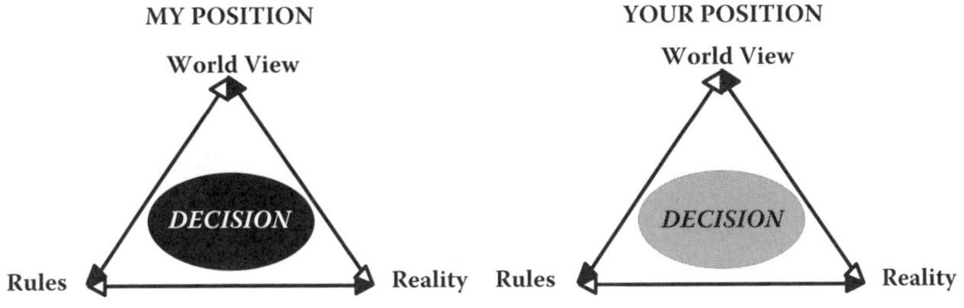

FIGURE 17.3
Conflicting Decision Maps as illustrated by Greg. (From Human Systems Dynamics, http://www.hsdinstitute.org/learnmore/library/faq-booklet.pdf. With permission.)

other leaders were following. All seemed to be paying close attention, with the small head nods that indicated they were tracking, so he continued.

"In this case, perhaps we could agree to abide by the same rule for getting our points across. It might be yours, it might be mine, it could be a hybrid of the two, or it could be something completely different. For example, we might agree—all of us, not just Warren—that if we want to be recognized to speak, we raise two or three fingers of our hand. However, if we're feeling really strongly, we raise our entire hand. By agreeing to change the rules element of the Decision Map, we might expect to see changes in our world view or pay attention to different data. We can't know in advance if the changes will be major or minor, positive or negative in terms of the particular conflict. There may be no change at all. However, it does change the pattern to increase the chance something more productive will occur. Is everyone here willing to try an experiment and use the fingers versus full hand for recognition?"

Mark Moore raised both of his hands high over his head, triggering laughter from the others in the room. "Can we take a break? My eyes are beginning to turn yellow." Everyone laughed and went on a short break.

18

Meanwhile, Back at the Projects ...

Staff meetings, training, and day-to-day operations continued. Participants from both teams attended the training classes together. Much to everyone's (except the consultants') surprise, the paths of the projects quickly diverged, each with their own challenges.

THE SHIP DATE TEAM

The "promised versus actual ship date" project's adventures started early. The team, staffed with a cross-functional, cross-facilities team of seven people, was eager to make something happen. Harry Helms, a midlevel manager who worked for Karen Decker in manufacturing, was identified as the team lead on the project. He was known as a go getter, ambitious, and smart. Harry was respected as a straight shooter and an excellent communicator with a pragmatic approach to work. During training, Harry took copious notes and asked lots of questions, mostly regarding the practical application of the concepts and tools being presented.

As the group learned about project charters in the Define phase, Harry took it upon himself to create a draft of his team's charter. His line of thinking was to come to the first meeting with something for the group to start with, figuring it would be faster to edit an existing document than create something from scratch. At the first team meeting, Harry opened the meeting with the usual greeting, agenda, and ground rules. Next, he distributed copies of the draft charter to the team members and asked for feedback. Harry, conscious of his ego, was prepared to be gracious in accepting the thanks of his teammates for helping them get started so quickly.

Instead Cathy, one of the team members, reacted, "What the heck is this? I thought we're supposed to be a team, with equal voices and input. I assumed this was a participative, democratic effort to address shipping times. Just because the leadership team anointed you Team Leader doesn't entitle you to make arbitrary decisions for the rest of us."

The thud of Harry's jaw hitting the table was audible in the back of the room. Greg Sanchez, the consultant coaching the team that day, noticed his nonverbal distress (and rising blood pressure) and decided to intervene. First Greg acknowledged Harry's contribution to the charter process and then asked him to give some background as to his rationale for drafting the document for the group.

Harry, feeling Greg's support, shared, "This project has a limited amount of time to generate results. I thought it would speed things up to document my understanding of the project as a starting point for team conversation. Nothing more, nothing less. Little did I know it would be an issue."

Greg thanked Harry and took over the conversation. "This is the perfect time to introduce an important HSD concept, the Eoyang CDE model.* It is tied to the concept of patterns, which we already introduced in our team orientation. The CDE model states that patterns can be described and influenced by understanding and shifting their Containers, Differences, and Exchanges.

"**Containers** hold the various elements of the system together; the way we organize or categorize ourselves in groups or clusters. There are always many different ways to characterize containers. You team members share the container of *THM* employees. Tina and I share the container of consultants. We have shared containers around gender, education, styles, and shared beliefs—lots of different ways. Any one person belongs to many different containers at the same time.

"**Differences** provide the potential for change within a system. Look at this rubber band dangling from the end of my finger. How likely is this system to change? If the status quo is what you desire, the relaxed band is probably just fine. But, what if you want, or need, the system to be different?" Greg then hooked a second finger through the rubber band and stretched it tight. "How much potential for change is there now in this system?" he asked, pointing the stretched rubber band at the team

---

* Human Systems Dynamics Web site, http://www.hsdinstitute.org/about-hsd/what-is-hsd/faq-about-the-hsd-institute.html.

members. Everyone laughed, wondering if he would actually shoot it at them.

Greg continued with his explanation of the CDE model. "**Exchanges** establish the connections between the various people and elements of the system. We use a variety of formal and informal, intended and unintended means to exchange information, goods, and services.

"Let's use the CDE model to describe the current state of the system as it relates to what just happened between the team and Harry. What was going on?"

Cathy spoke up, "I guess I thought we were all in the container of equal team members. When Harry drafted the charter independently, when he acted differently, I took that to mean he saw himself in a separate, (bigger and better) container of Team Leader. Reacting to that difference, I perceived his exchange of presenting the draft charter as a one-way message and reacted accordingly."

The other team members nodded their heads in agreement. Harry responded, "Wow! I didn't realize all of that was going on. I don't think of myself as separate from the team in a separate container. In fact, if someone else was named Team Leader, I would have still drafted the charter so we could move more quickly. My intent was that this was a starting point, a draft, expecting the differences to be made bigger by the group's input. I never wanted this to be a 'my way or the highway' conversation, rather the opening up of a two-way exchange."

Greg applauded, clapping his hands in acknowledgment of the two skillful conversations that had just taken place. "We'll continue to use the CDE model as a descriptive tool. In fact, let me show you how it can help with project definition."

On a flipchart, he drew a simple table with three columns. The first one identified the departments in the room. The second provided a space to define the differences between the departments as they related to shipping, and the third to list the kinds of information needed for that department to act. The blank table is illustrated in Figure 18.1.

Greg then asked the team to complete the chart to understand what each department did along the shipping process that was different and added value to the process. The team went straight to work and completed the chart, as seen in Figure 18.2. Greg asked what patterns the team members noticed. Several were identified, including:

The Ship Date Process

Departments	Difference They Make to the Shipping Process	Kind of Information Needed for That Respective Department
Sales		
Production scheduling		
Manufacturing		
Quality		
Shipping		
Billing		

FIGURE 18.1
Blank Ship Date Team current process form.

- Each department is primarily concerned with their own responsibilities—getting their own work done.
- The customer's requested ship date gets lost about halfway into the process.
- There is clear accountability for the order throughout its life cycle.

By completing the chart, the issues that needed to be addressed emerged for the Ship Date Team. Greg asked Harry to share his draft charter with the group once again to compare what was drafted to what was now understood to be the challenge. With a few minor revisions, the team agreed the charter was now "good to go," so much so that Cathy ended up thanking Harry for providing the charter.

Harry asked Greg about the connection between the chart they had completed and the CDE model. Greg responded by asking the group to make the linkage. The discussion highlighted the idea that the first column identified the various departments (containers), the second explored the significant differences as they related to their project, and the third, the meaningful interactions between them (exchanges). "By understanding how each of the various departments acts to achieve the shipping process, we can then see the patterns that emerge," commented Greg.

Greg paused, thinking. After a couple of seconds, he refocused his attention on the group. "I'd like to introduce another HSD concept called *Simple Rules*. As this team progresses, it might be valuable for you to establish a list of simple rules for yourselves. Creating them helps surface the 'differences that make a difference' and can provide some confidence

The Current Ship Date Process

Departments	Difference They Make to the Shipping Process	Kind of Information Needed for That Respective Department
Sales	Get the order; create the demand	Quantity Type Requested delivery date
Production scheduling	Schedule the parts for production	Customer's requested date Knowledge of standard lead time Status of relevant manufacturing areas
Manufacturing	Build customer requested products	Customer's requested date Current backlog Anticipated outages/shortages Standard lead times Actual cycle times Customer specifications
Quality	Assure products are internally and externally compliant	Customer specifications *THM* specifications Known variations Historical issues
Shipping	Pack and ship the product with appropriate paperwork	Expected quantity Packaging requirements Relevant regulations
Billing	Bill for products; reconcile any discrepancies	Quantity Date shipped Anything out of the ordinary

FIGURE 18.2
Completed Ship Date Team Containers, Differences, and Exchanges (CDE) application exercise.

that everyone will have the flexibility to do what he or she needs to do to adapt to unexpected changes during the project. Simple rules should not to be confused with team norms or simple working agreements, such as 'start and stop meetings on time.' Simple rules create a general framework that helps explain how to operate within the team or organizational system. They help people understand how to operate and move more quickly within the system toward the team's mission, goal, or objective. Typically, only five to seven of these are needed and they usually include agreements about 'containers,' 'differences,' and 'exchanges' among them." Below are the simple rules created and used by the HSD Institute:

- Teach and learn in every action
- Reinforce strengths of self and other
- Give and get value for value
- Attend to the part, the whole, and the greater whole

Greg continued, "Tina and I went through this process ourselves as we figured out how we would need to work with each other and with you. Here are the rules we created for ourselves."

- Align our efforts for the highest good of the project
- Provide seamless support for our respective disciplines
- Teach to provide value in every interaction
- Model learning by always learning from the client and each other

Greg moved to the whiteboard. "It seems like it might be useful to create a simple rule regarding team leadership and decision making (a *difference* and *exchange* rule)." Greg led the group through a brainstorming session, emphasizing that simple rules have to do with behaviors as opposed to values, which he characterized as dealing with internal emotional states and agreements. After some conversation about the brainstormed ideas, the team settled on the following simple rule: Make timely decisions, incorporating the input of other team members whenever feasible. Greg pointed out how the rule created a space big enough for both Harry and Cathy's world views regarding leadership.

"Here are three other thoughts about simple rules before we continue our rule generation exercise," said Greg. "First, everyone in the system, regardless of rank, tenure, or anything else must agree to follow these rules. Imagine rush hour traffic where one driver gets it into his or her head to drive on the wrong side of the road. It is essential that team members trust that everyone will abide by these rules; it's one of the reasons I'm investing the time to have you build them together. Second, the rules should be 'robust.' They should hold true over a wide variety of different situations and over a long period of time. Third, the rules might not always be in perfect alignment with one another. That's okay; it simply alerts you as a team that this is an area of difference. Remember, differences in a complex system provide the potential for growth and self-organization. Any questions?" The team continued on to develop the remainder of their simple rules and the meeting continued.

THE ACCOUNTS RECEIVABLE VARIATION (ARV) TEAM

Meanwhile, the "reducing variation in the accounts receivable for invoices coded net 30 days" team learned a different set of lessons. While they did not experience a team leadership issue, they did bump up against a different concern. Kathleen, the ARV team leader, asked what she thought was a fairly straightforward question. "We're going to go after the biggest offenders first, right? We can use the Pareto principle—the 80/20 rule, to find the 20% of the offenders that cause 80% of the late receivables."

Frank, another team member who was part of the sales organization, raised his hand. "I'm more than a little concerned about the political and long-term business implications of that decision. Some of our biggest customers are in that 20%. In this marketplace, getting them angry at us—especially when our delivery performance hasn't been the best—seems like a significant business risk. If they leave us, the financial impact will be far worse than the loss of 15 days of their cash." Heads nodded around the table.

Tina Grice, who was coaching the team that day, applauded. "This is a *great* conversation. Kathleen's application of the Pareto principle was right on target. Frank's concern about other factors—the bigger container—is also spot on. Said differently, here is an example of what Greg and I have been discussing during the training, a situation where linear models meet complex adaptive systems. It's not a matter of right or wrong, rather, it is an opportunity to determine how to merge these two different paradigms." The team was quiet, digesting Tina's comments and contemplating how to move forward.

Tina continued. "Let me share a new concept in hopes that it might create some new opportunities for action. Has anyone ever heard of fractals?" Eva raised her hand. "I think they have something to do with being 'copies within copies,' like a head of broccoli. Is that what you're talking about?"

Tina nodded her head. "You got it." she said. "Fractals are a property of certain nonlinear systems. Complex adaptive systems exhibit layers of order and chaos in time, scale, and space. Self-similarity—the ability to maintain the same general shape at many different scales—is one of the defining characteristics of fractals. Fractals appear in mathematical models. Let me show you a picture from one of the most famous, the Mandelbrot Set (Figure 18.3). She showed the slide on the screen.

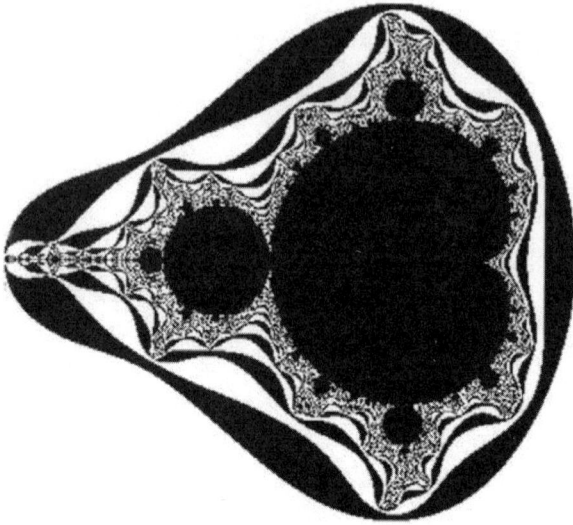

FIGURE 18.3
The Mandelbrot fractal. (Illustrated by Lois Lenox.)

"You can see how the smaller 'nodes' are similar in shape to the larger ones. This remains true as any one area is magnified."

Tina smiled. "You're probably thinking to yourself, 'This is interesting, but what does it have to with our ARV problem?'" Several heads nodded affirmatively. "I've got a candy bar for the first person willing to volunteer an answer." Even with the promise of chocolate, there were no responses. "Let me give you a hint," said Tina. "Are all accounts receivable the same?"

Kathleen, true to her outgoing form, raised her hand. "I'm hungry and could use some candy right now," she said. "I think you're hinting that late accounts receivable is a fractal, that there is a consistent pattern regardless of size of customer. Is that it?"

Tina tossed the candy bar to Kathleen. "Perfect." she said. "Now, for another candy bar, what new possibilities for action does that idea trigger for you?"

Frank, who had been listening intently the entire time, cleared his throat. "One possibility is that we should try to influence the pattern on small customers. If we screw them up, it won't cause as much damage." Frank leaned forward in his chair, beginning to get excited as he continued to think out loud. "And once we get it right with the small guys, if it is indeed a fractal, then it should work in a similar way with customers of all sizes, right?" A second candy bar sailed through the air in Frank's general direction.

Throughout the next several months, similar issues and concerns surfaced for both teams. With the coaches there to assist, each team worked its way through the intricacies of their respective issue, applying the appropriate Six Sigma or HSD tool as needed, and sometimes a combination of both. Greg and Tina took the role of coach seriously, staying on the sidelines and trying to ask questions that would lead the participants to their own answers. When needed, they introduced new tools, taking advantage of the "teachable moments" that inevitably arose.

A new simple rule emerged out of the conversations that Tina and Greg had with each other, reinforcing for both of them that Six Sigma implementation was a fractal in its own right. After receiving some feedback that suggested some were learning tools that others weren't, the two instructors agreed that a new simple rule would be to introduce and teach *all* tools to everyone. Even if the tool wasn't immediately useful to a given team or situation, it would add tools to everyone's toolbox for possible use in the future. To do this, Tina and Greg agreed at the beginning of the meeting, one of the first agenda items would be to update the respective teams with the progress of the other team. The discussion would highlight any new tool introduced to the other team. In addition, and if possible, a member from the other team would be asked to join them for that portion of the meeting. That person's role would be to present the other team's process improvement challenge, the new tool the team used, how it was applied, the new insights it created, and, lastly, the results that using the tool achieved.

Four months later, both project teams had worked their way through the Define, Measure, and Analyze phases and were implementing their pilot solutions. Initial feedback was positive. Six Sigma appeared to be off to a healthy start at *TryinHard Marine*.

19

One Year Later

It had been about a year since *TryinHard Marine* had embarked on its Six Sigma journey. John Saylor and Tamara Hart were relaxing over lunch at a CEO Roundtable meeting. They had conferred early and often over the past year, with John more and more comfortable asking Tamara about her Six Sigma implementation experience. Tamara, for her part, found the opportunity to step back and reflect extremely valuable. It was an opportunity to conduct a series of informal "after-action reviews"; a chance to maximize her learning by discussing what had worked well and what she would have done differently if given the chance for a do over.

"So, John, is *THM* where you thought you would be with regards to Six Sigma at this point in time? As I recall, you were concerned about a repeat of your TQM implementation experience."

John smiled as he recalled their initial conversation. "To paraphrase an old commercial, 'We've come a long way, baby,'" said John. "We're actually farther along than I expected. I would have been satisfied if we had completed our initial projects, recouped the cost of the consultants, and didn't have people shudder whenever the term 'Six Sigma' was mentioned. In fact, we've done those things and more."

John leaned forward, absent-mindedly counting the accomplishments on his fingers as he listed what he saw as the current state of Six Sigma at *THM*:

- **Different conversations taking place.** Instead of the usual finger pointing, blame seeking behavior that was typical whenever something went wrong, there was an increasing use of a different question: What in our process allowed this to happen?
- **Self-organization around process improvement.** In the past, employees at *THM* would identify issues, then stand back and wait

for someone in management to "fix it" for them. There was a lot of "That's not my job" invoked at all levels of the company. Today, John noticed a (slowly) increasing amount of self-organizing—one employee (often not a Black Belt) saying to another, "Would you be willing to work with me on this?"

- **Leadership is still talking about it.** In particular, *THM*'s TQM implementation had been characterized by its usual pattern of a big, splashy kickoff, then the leadership team moving on to "the next big thing." Building Six Sigma progress into the regular fabric of the staff meetings helped keep the subject "top of mind" for the leadership team. Having the consultants make regular "guest appearances" also helped create a sense of accountability.

- **Built into budgets, performance expectations, and incentive compensation.** This was another reason (actually a *big* reason) the leadership team was so engaged. John found himself paying closer attention knowing that 30% of his bonus was predicated on making progress on a "balanced scorecard" of Six Sigma implementation indicators. Creating that scorecard had been a challenge in its own right: balancing "hard" and "soft," short-term results and long-term investments; "catching the fish" results, and "teaching the organization to fish" process and patterns. It had been worth it, though; it created constructive conversation among the leadership team about the trade-offs involved as they made decisions that affected not only Six Sigma, but other aspects of organizational strategy as well.

- **Performance measures are transparent and steadily improving.** Another benefit of the emphasis on "both/and" results—the shorthand phrase that Mark Moore had coined to capture the "dynamic tension" between the different measures—was that the entire organization had a better sense of the overall status of the business. The consultants had challenged the leadership team to make the balanced scorecard visible to all employees ("Is there anything here you're ashamed of, or think needs to be kept secret from the only ones who can make these numbers change?"), and after some misgivings, they had agreed. Each vice president shared the results as part of his or her monthly staff meeting, and the results had been dramatic. The employees asked great questions, often pointing out inconsistencies and opportunities they noticed. An unintended consequence was that the vice presidents didn't want to look ignorant in front of their people, so they quickly learned to make sure they really understood the numbers themselves.

- **Customers notice.** Much to Brian Rudd's delight, his expectations in supporting Six Sigma implementation had not only been met, but exceeded. Customers *had* noticed the changes in *THM*. They certainly noticed the improved delivery ship dates. These improvements, brought about by one of the pilot Six Sigma teams, had resulted in several new orders. The "exceeded expectations" came from a subtle shift in the tone of the conversations his customer service representatives (CSRs) had with customers. While this group received little formal training in Six Sigma, somehow the importance of listening to customers had "seeped into their pores through osmosis," to quote Brian.
- *THM's* **culture is perceived as different.** This "seeping into the pores" behavior was not confined to the CSRs. While not as obvious, there did seem to be a different feeling when walking in the door of *TryinHard Marine*. The differences were subtle, to be sure: different questions being asked, a little more pep in their step, more people talking to different people than ever before.

John laughed, leaning back into his chair and shaking his head as he looked down at the seven fingers he was now waving in the air. "Not bad," he said, "especially compared to the number of fingers I would have raised if asked this question a year into our TQM implementation."

"Well, John, it would seem that my days as your 'shadow coach' are over. It's been a pleasure to watch you grow personally as well as to enjoy the success of *THM*. I have two other, I hesitate to say 'final' because I hope we'll continue to compare notes about this, thoughts for you."

"I'm all ears," said John.

"My first thought—borne from my own experience—is to not get complacent. *THM* has more experience with its old patterns than its new one. You and your staff will need to continue to nurture the spark you've created, otherwise you're at risk of claiming victory and moving on to something else. Trust me; I learned this one the hard way." John nodded, understanding how easy that would be to do.

"Second, continue being a student of your own patterns, the patterns of your leadership team, and the patterns of your organization. The issues we face, the 'fish we need to catch' change daily, especially in complex systems. I've found it much more valuable to focus attention on the 'learn to fish' skills; the pattern recognition and influencing skills. They are the transferable ones, the 'gifts that can keep on giving.' I'm guessing that Greg Sanchez spoke with you about the fractal properties of certain complex systems. I've found it extraordinarily useful to remember that, in fractals, a small part of the system replicates the basic shape and properties of the larger system of which it is a part. Each of your CSRs **is** *TryinHard Marine* to your external customers. Each of your first-line supervisors is *TryinHard Marine* to your hourly employees. You embody, whether you think you do or not, *TryinHard Marine* to every other CEO at the CEO roundtable meetings."

John nodded his head slowly. "I guess I hadn't thought of fractals in organizations before. I've read in a dozen management books about how leaders need to be role models for their organizations. Thinking about leaders as fractals puts a different spin on the role model thing."

This time it was Tamara's turn to nod. "Fractals have another implication that I think is counter-intuitive to most contemporary leadership books. How many times have you read, 'All change must be driven from the top-down?' Fractals imply that change can happen anywhere in an organization and if the conditions are right, can ripple up, down, and sideways to influence the larger system."

John responded. "I hadn't thought about fractals that way, either. What I think you're saying is that we need to encourage innovation, learning, 'try an intelligent experiment' thinking everywhere in *TryinHard Marine* because we don't know which of those 'experiments' will bear fruit."

Tamara answered, "That's what I'm trying to do in my organization. It's a slow process, or at least that's how it feels right now. I'm having fun, though, and I think it's good for the morale of the organization to know they are 'empowered.' I use that word cautiously at *Stars in Health* (it has such a buzzword connotation) to make change and improvement happen."

John hoisted his glass in a toast. "Here's to continuous learning and continuous improvement in both of our organizations." They touched glasses and enjoyed their toast.

QUESTIONS FOR CONSIDERATION

As we did at the end of the first case story of *THM*, please reflect on the following questions, either individually or with a group of colleagues:

- Does the *TryinHard Marine* Take 2 case story resonate with you? For what reasons?
- What was similar in the way *TryinHard Marine* implemented Six Sigma to your business experience? What was different?
- Did *THM* achieve the ability to be agile to changing conditions? Why or why not?
- How well was alignment and buy-in achieved by *TryinHard Marine*? What made a difference?
- Based on the case story, do you believe the *THM* implementation will be sustainable over time? Why or why not?
- Of the models or tools presented in the second telling of *TryinHard Marine*, which ones resonated with you and why?

Section IV

Now What?

20

Now What?

We have traveled quite a distance through this book. The *"What?"* section provided a historical overview of the development of three parallel patterns. The overviews of Lean enterprise and Six Sigma in Chapters 1 and 2 described a deep pattern almost 100 years in the making. This pattern—linear, cause and effect processes, and systems that can be optimized to the one best way—was an essential driver of the industrial revolution. One of the implications of that world view is people—as individuals, groups, and organizations—can also be viewed through this mechanistic lens.

Almost simultaneously, and based in part on the pervasiveness of the idea that organizations should run like well-oiled machines, a parallel exploration took place. Focusing on the human side of processes and change, Chapter 3 provided a brief recounting of some the major milestones in human relations, organizational behavior, leadership, and organizational development. This timeline culminated in the 1990s with the popularization of systems theory, deterministic chaos, and complexity theory. These are some of the sciences that underlie Human Systems Dynamics (HSD) described in Chapter 4, and recognize that human systems are complex, self-organizing, and better explained by patterns than by cause and effect. This paradigm shift is the reason for this book. It explains the reasons many process improvement initiatives have not lived up to their potential. Through its different way of viewing the world, HSD offers new, practical possibilities for action.

In the *"So What?"* section, we utilized the convention of a case story to highlight two different ways of understanding the dynamics that occur when implementing linear process improvement concepts and tools. In the first telling of the *TryinHard Marine* experience, we exaggerated (though based on some of our client experiences, not too much) some of the dynamics that underlie the worst of a traditional, linear Six Sigma deployment. Leaders make decisions to go forward with incomplete or inaccurate information. Teams are under pressure to "move through the steps" and "put checks in the boxes" of training and project completion. Teams fail to account for the differences and changing conditions that are an inevitable part of organizational life. Blaming and finger pointing is common; suboptimal results accrue in both the short and long term.

In the "*So What—Take 2?*" section of the book, we revisited the *THM* case, this time with the addition of Human Systems Dynamics. Tamara Hart fulfilled the role of "sage mentor" to the CEO. Greg Sanchez provided the "voice" and sensibilities of HSD to the projects. Several of the same scenarios described in the first telling of the story were repeated to provide a "same and different" comparison when the principles of HSD were considered. Some new scenarios were described, recognizing that adding a complex adaptive systems paradigm to a strongly held linear paradigm creates its own set of issues, even as it eliminates others.

All of which brings us to this final section of the book. We hope you're thinking something like, "Okay, I'm intrigued. The addition of these new concepts and tools offers potential new ways to rethink our past/present/anticipated (circle the appropriate ones for you) approach to process improvement. The problem is, based on the case story, I'm not sure I understand how to go about doing it."

Thanks for asking. This section will provide a high level flow (to give a detailed roadmap is impossible; too many variables in a complex system, remember?) that can (and will have to) be adapted to your individual situation.

Note: To repeat a caveat offered earlier, "*What Works for GE May Not Work for You*" is not a textbook. Others have already written detailed manuals for how to use Lean, Six Sigma, and HSD concepts and tools. We have mentioned some of them earlier in the book and list others in the Recommended Reading section at the end. Refer to them early and often as needed.

Another Note: Six different aspects of an enhanced process improvement implementation pattern (Figure 20.1) will be described. We deliberately do not number them. Doing so creates the sense there is a strong, linear, step-by-step sequence to them, and it isn't so. Each element is connected

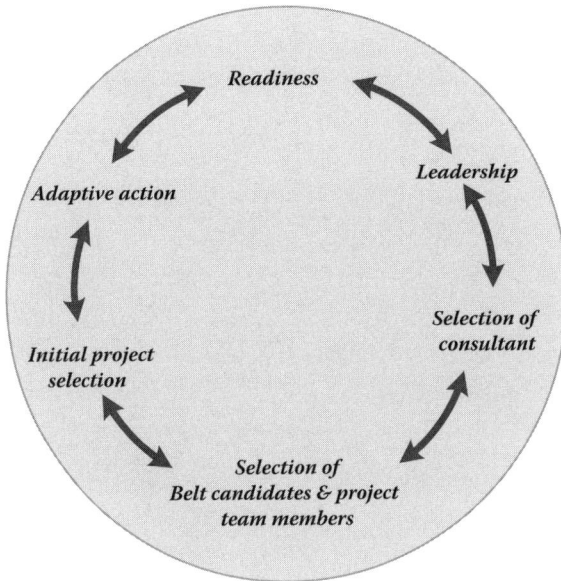

FIGURE 20.1
Process improvement implementation pattern.

to every other one, "semiautonomous agents, each working independently, that self-organize to create system-wide patterns." Even with this caveat, describing each part is valuable. It provides shared language, questions to ask, and tools to consider when trying to positively influence change.

READINESS

Unless your company is a new start-up enterprise, patterns relevant to process improvement already exist, and they are rooted in each individual's past experiences and the organization's collective story and culture. It has been said that, "Those who ignore the past are doomed to repeat it." We recommend you make a concerted effort to collect answers from a variety of perspectives within your company to questions, such as:

- What process improvement (PI) models and tools have been introduced in your organization?

- How long ago?
- How were they implemented?
- What results did they achieve? How were those results measured?
- How much of what was implemented then is still in practice today?
- What worked well? What didn't work well?
- What is the same today in the culture as it was the last time PI was implemented? What is different today?

The answers to these questions will provide valuable inputs to a Same and Different analysis, or a CDE (containers, differences, exchanges) analysis of organizational readiness.

Another aspect of readiness is need and motivation. The authors would be a lot closer to retirement if we were paid a dollar for each time someone uttered the phrase: *"process improvement program." "Program of the month," "This too shall pass," "Why fix what isn't broken?"* The program pattern in many organizations is very strong. It is interpreted as meaning discretionary, temporary, and unrelated to real work. Instead, we recommend that the leadership team take a clear-eyed, honest look at their motivation for implementing process improvement in their organization. Some questions to ask:

- How, specifically, does this help us accomplish our mission?
- How will we convince skeptical employees that we're serious about this?
- What commitments are we, individually and as a leadership team, prepared to hold ourselves and each other accountable for?
- What are the consequences of a successful change to a process improvement culture? How will we measure them?
- What are the consequences of a failed change? How will we measure these?
- Why now? What is gained or lost by waiting for another time?

A third dimension of readiness is connection to larger systems and patterns. Context sets the initial conditions for a pattern. Chaos theory suggests that complex systems are very sensitive to initial conditions, that small initial changes in those conditions can create large differences over time. Given these considerations, questions to ask include:

- What is the economic environment today? What changes are expected? What difference would it make to our proposed process improvement (PI) implementation if they occurred?
- What is the competitive environment today? What changes are expected? What difference would it make to PI implementation if they occurred?
- What is the technological environment today? What changes are expected? What difference would it make to PI implementation if they occurred?
- What is the current state of our employees today? What is their level of readiness and willingness to engage in the additional work that often accompanies a change in pattern?

LEADERSHIP

We ask you to interpret leadership broadly. Obviously, job descriptions with the words *chief* _____ *officer, president, vice president,* and *director* imply positions that directly or indirectly influence others in the organization. Recognize that many others in the organization also exert influence in formal and/or informal ways. These *influencers, gatekeepers, networkers,* or whatever else they might be called, generate the same fractal properties as the CEO.

Leaders have a critical role in establishing and influencing organizational patterns. That role begins with an honest assessment of their individual patterns: strengths, weaknesses, prejudices, and blind spots. The fractal nature of many complex systems suggests that "Do as I say, not as I do" is unlikely to work. In fact, the self-similar and scale-free properties of fractals imply that individual reactions to a change in pattern are likely to be replicated throughout the organization. In the *TryinHard Marine* story, John Sailor had to confront his own patterns based on Tamara's gentle coaching. Only after he accepted the challenge to change his own pattern could he stimulate similar changes in his staff using the "past and future commitments to PI success" exercise.

Asking the following kinds of questions will help create the powerful conversations needed to assess and influence this element of a process improvement pattern:

- How willing are the leaders to change their own behavior?
- What are the "sacred cows," the behaviors or patterns they are unwilling to change?
- How comfortable is the leadership team with dealing with ambiguity? With embarking on a change that has no guarantee of success?
- How willing is the leadership team to deal with conflict as it pertains to changing the pattern? Is this a "battlefield they are willing to die on?" Are leaders willing to provoke conflict to amplify differences in support of pattern change?
- What consequences are leaders willing to bring to bear to reward success or address failure? Are they willing to put their own and others' financial "skin" in the game?
- Are the leaders willing to do the hard work of original thinking? Are they expecting a "plug and play" solution from a textbook or consultant? Are they willing to integrate and synthesize ideas and concepts from multiple sources—or create brand new solutions—in service of customizing the pattern change to the organization's unique situation and context?

The last question bears reinforcing. Lean Six Sigma—or any continuous improvement initiative for that matter—is not a one size fits all program to be spoon fed to the organization. These efforts require active leadership at multiple levels of the organization.

SELECTING A CONSULTANT

Selecting a partner or partners to help with process improvement offers many potential benefits. They include:

- **Changing the Container.** Adding new people to the mix breaks down cliques and offers new opportunities for grouping members of the organization together through training, project teams, etc.
- **Amplifying Differences**. Effective consultants bring their experiences from other clients, geographies, and industry segments to

a particular engagement. Their sharing of stories, questions, best practices, and lessons learned can help stimulate new thinking and create the tension needed to change existing patterns. *Note to consultants*: Tread carefully here. On one hand, too many stories too soon can trigger a defensive, "Not invented here" or "That doesn't apply to us" response by clients deeply rooted in their own patterns. On the other hand, the judicious use of stories to illustrate how other organizations used these new concepts and tools to change their patterns, provides illustrations of successful adaptive change management.

- **Change exchanges**. Through the vehicles of training and individual, team, and executive coaching, effective consultants influence the frequency, channel, and content of the conversations and stories being told in the organization.

The above bulleted list describes the differences the addition of the consultants make to the organizational pattern. Effective consultants also possess the tools to intentionally influence **C**s, **D**s, and **E**s within the organization to help move it in desired ways within the system landscape. Figure. 20.2 illustrates the range of choices available.

As highlighted in the case story, there are three different competencies internal or external consultants need to possess and use. There are

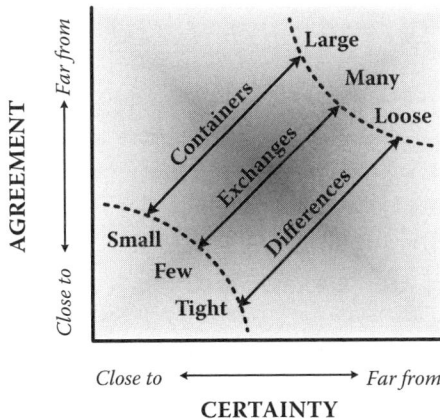

FIGURE 20.2
Adjusting conditions for self-organizing. (With permission of Human Systems Dynamics, Inc.)

occasions where one person or firm will have strengths in all three of the areas listed below, and others where it might make sense to engage multiple resources to be sure all support functions are provided with quality.

- **Process improvement concepts and tools.** We don't believe that people can learn Lean or Six Sigma strictly by reading books on the subjects. Effective consultants translate technical/statistical tools and data into language the organization can understand and use. Knowing what tools to use when, how to modify them, their "tips, traps, and failure modes," each of these can prevent unnecessary heartache.
- **Human Systems Dynamics concepts and tools.** This set of competencies mirrors the one above. Because the field is still emerging, there is less written about it; a smaller body of knowledge and experience exists. This places more of a demand on the consultant to be knowledgeable and conversant in both HSD theory and application. Because it emphasizes patterns, deep industry or functional subject matter expertise is less important. These consultants need to be extraordinarily adaptive, agile, and creative.
- **Process consultation.** A great deal of process improvement work happens within a portfolio of individual projects. Each project has its own team that does its work in an even smaller container of individual meetings. Each of these—the portfolio of projects, the individual teams, and specific meetings—has its own set of dynamics, its own patterns.

Effective consultants serve as facilitators, or as Larry Solow says, "boxing referees." His explanation: "A boxing referee has several functions. The first is to keep all of the combatants safe from harm. The second role is to assure the fighters fight fair; following the rules that were mutually agreed upon before the fight began. Third, the referee assures that the participants engage, do the work they were brought there to do and not shy away from their responsibilities—even if they might be painful." Effective consultants do exactly the same thing in the context of process improvement projects, meetings, and teams. They create a space where differences can surface—bring conversations "to the edge of chaos"—in order to provoke new thinking and possibilities for action.

The following questions may be helpful for a leader to ask when selecting internal as well as external resources:

- What is your academic/educational background in each of the three areas above?
- What is your related job experience in each of those three areas?
- What are the results you've achieved in each area?
- Have you partnered with other internal or external change agents on a project? Please describe what worked well and not so well in that collaboration.
- How would you describe your ability to be agile and flexible? Give an example of a time you demonstrated this due to a real-time change in your client's situation.
- How will you measure your success?
- Where else have you delivered a project similar to the one we are discussing with you? May we contact them?

SELECTING AND TRAINING BLACK BELT CANDIDATES AND PROJECT TEAM MEMBERS

Highlighted in the case story, the selection and training of team members and Black or Green Belts constitutes a major investment by the organization. Some firms consider their Belts, especially their Black Belts, as internal consultants. This means the three competencies listed for selecting a consultant: technical, HSD, and process consultation skills become areas to assess for potential BB candidates.

Some other criteria to consider during team member or "Belt" candidate selection include:

- Broad knowledge of the organization
- Formal and informal network of colleagues
- Perceived credibility by others in the organization
- Commitment to the human side of project management
- Change management skills
- Superior communication skills
- Ability to deal with ambiguity, the things that don't go according to plan
- Ability to "herd cats"; deal with busy people who have differing agendas and competing commitments

- The ability to keep their ego in check and operate with a level of humility. They need to recognize that the BB job is no better or worse than any other and, in fact, depends on other employees to be successful.

Trying to identify candidates who already possess every one of these skills and abilities in abundance is often difficult. The good news is that it is not essential that every person hit the ground running with every item on the list fully developed. The organizational challenge is first to create and implement assessment tools that define a candidate's current skill level in any given criteria. The second challenge is to create viable plans to develop those skills, along with an evaluation process to determine the amount of progress being made.

INITIAL PROJECT SELECTION

One pattern we notice in many clients is the tendency to over reach in their initial selection of projects. While understandable ("I need a return on my investment as quickly as possible" or "Let's work the worst area … anything we do will help it"), this rationale often sets up a negative chain reaction (Figure 20.3).

The way to break this pattern is to be very conscious of project selection criteria and setting the initial conditions in such a way that their probability of success is maximized. Some ideas to consider include:

- Are we ready yet for this project?
- Will the leaders of those projects actually lead?
- Which consultant will be the best fit for the project?
- Can we get a "quick win" (versus a long, drawn out timeframe)?
- Would solving this issue be popular with the employees, management, and staff (versus one that might result in layoffs, additional hardship, or grievances)?
- Is it aligned with the current goals, objectives, and priorities of the business (versus a training or make-work project)?

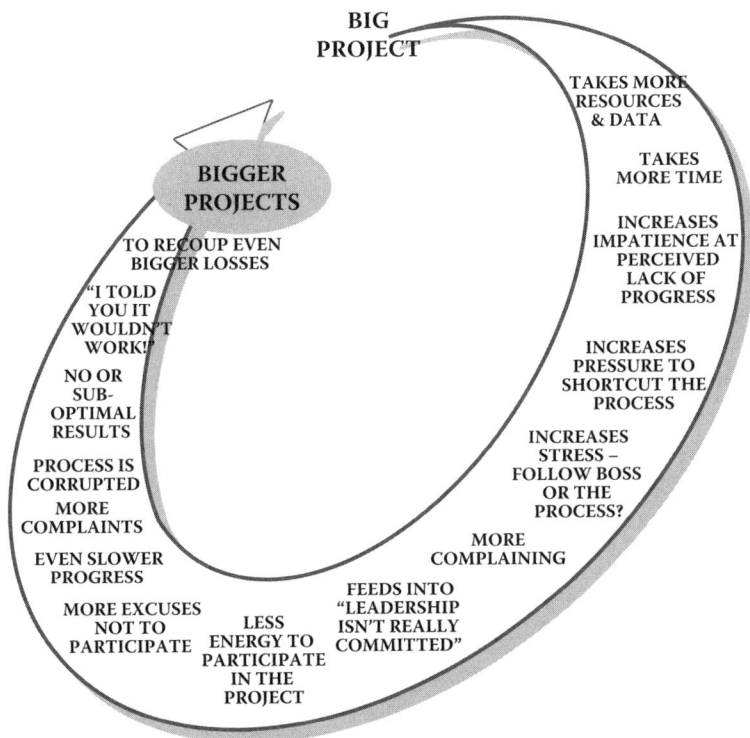

FIGURE 20.3
Project "spiral of doom" pattern.

- Would solving this issue lead to a natural progression of subsequent projects (versus a "once and done" effort)?
- Does data currently exist that might be usable and useful (versus data does not exist, is not of acceptable quality, or is not relevant to the specific project's needs)?
- Is this issue fairly well self-contained (versus touches multiple processes, functions, or departments)?
- Is it within the control of the team to solve (versus requiring customer or governmental approvals)?
- Can the results/benefits of the project be easily measured (versus a project to change attitudes or avoid costs)?
- Are the resources available to implement and maintain the solution once it is identified (versus "That's a great idea, but we can't afford it.")?

ADAPTIVE ACTION

The organization is ready. Leadership is onboard. Consultants, team members, and initial projects have been identified. Whew! Now all that needs to happen is to implement, right? We wish it were true. It is not likely to happen. Complex systems are continuously changing; they have to in order to survive and flourish. Conditions change continuously. Many of these changes will have no immediate bearing on the process improvement implementation, some will only impact it later, and others not at all. Other transformations will have immediate impact, sometimes out of proportion to their size.

What's an organization to do? Adapt. Figure 20.4 illustrates the Adaptive Action cycle based on the questions of "What?", "So What?", and "Now What?"

As the work begins and continues, savvy observers of these organizational patterns will constantly be describing what is currently happening (take off your rose-colored glasses for this part) to the current organizational pattern. Once the current state has been identified, evaluation takes place to determine if "where we are now is where we want to be." If the answer is no, then the identification and prioritization of potential changes can take place (the "So What?"). Once identified and prioritized, implementing the change interventions influence the pattern (the "Now What?") and the cycle repeats.

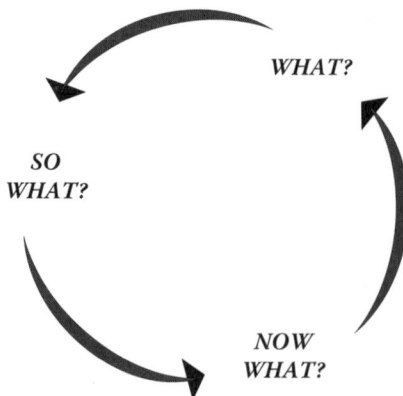

FIGURE 20.4
Adaptive Action model. (With permission of Human Systems Dynamics, Inc.)

The various HSD models and tools introduced earlier in the book provide specific distinctions and language that can help in each of the adaptive action steps.

The **CDE Model** invites questions.

Container questions include:

- Who are the various groups relevant to the process? Do they include external customers?
- Are cliques being formed for unknown reasons?
- Are the boundaries between containers becoming more porous? Is information and energy being transferred differently?

Differences questions include:

- What new distinctions are being noticed? Are differences in variation, types of waste, different patterns, etc. being identified?
- What rewards are being given by whom and to whom?
- How are differences being measured? Are those measures more or less sensitive than they were previously?

Exchange questions include:

- How much of what type of information is being shared?
- Are people talking and listening to different people than they were previously?
- Who is included in decision making and action?
- Is there a change in the amount of one-way telling versus two-way conversing?
- How well are (the inevitable) rumors being addressed?
- How is technology being used to enhance communication and exchange?

The **system landscape** (Figure 20.5) provides another descriptive lens:

- What are the levels of certainty and agreement regarding key properties and behaviors of the processes and patterns under review?
- Is this a good "fit" with the organization's current situation? Future situation?

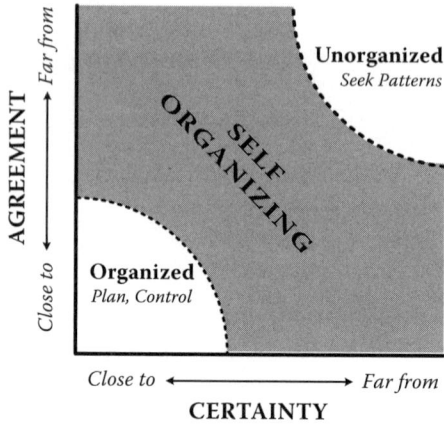

FIGURE 20.5
Ralph Stacey's and Dr. Glenda Eoyang's Landscape Diagram. (With permission of Human Systems Dynamics, Inc.)

The **7 Cs Matrix** (Figure 20.6) identifies another set of distinctions that can be used to assess, prioritize, and intervene to influence organizational patterns.

If all of this sounds like a lot of work, you're right. It would be so much easier if teams, organizations, community, and surrounding environment would just follow the plan we so carefully put together. By this point, though, we hope you have come to the same conclusion we did: wishing it wasn't so doesn't make this new reality go away.

Instead, we invite you to embrace this new paradigm while not discarding the old one. To repeat: linear, mechanistic models are incredibly valuable in the appropriate situation. In the highly controlled space of the Landscape Diagram, minimizing variation is essential. However, there are other regions in the landscape where the application of this mechanistic paradigm is not effective. With its new language, tools, and possibilities for action, HSD offers additional ways to positively influence change in complex, rapidly changing environments.

This is the challenge: to embrace the tension of these two paradigms. Brenda Fake talks about this tension as "functioning simultaneously in two worlds." With recognition that there are no guaranteed fixes, change agents are able to bring badly needed new tools and thinking to the organizations we serve.

Dimension	"What?"	"So What?" Interventions Include	"Now What?" Interventions Include
Connections	How do we do this today?	How well does this work now?	What can we add? What can we
Communications	What are the different forms it	How well will this work in the future if	subtract? What difference can
Continuous assessment	takes?	conditions change? Where is there most receptivity?	we amplify? What difference can we minimize?
Commitment		Which has the greatest ability to	
Coherence		positively influence others?	
Constructs			
Capacity			

FIGURE 20.6
7Cs and Adaptive Action matrix. (With permission of Human Systems Dynamics, Inc.)

In the Section I Summary, we closed with an interview with a "real" company that has experienced first hand the addition of the HSD paradigm to their traditional process improvement strategy. Focusing on the description of their current state, they described some of the "differences that made a difference" in their pre-HSD Six Sigma experience.

We close this section picking up with the second half of those interviews. Notice the experience and learning of "real world practitioners" as they incorporated some of the lessons and tools of HSD to influence their own work and increased their understanding of adaptive change management. (BF: Brenda Fake, RF: Ron Fischer, MK: Markey Keating.)

BF: Let's move to the next question. What was the rationale for adding adaptive change management to your process improvement efforts? What has been your experience to that addition in terms of the work delivered?

RF: We needed a new perspective that could help set conditions where a change management focus and skill set could be added because there was a big gap to help stuff (changes) stick. This is the key to achieving sustainable results. Six Sigma preached about achieving sustainable results, but spent most of the time teaching technical skills. This became a self-fulfilling prophecy because the people teaching did not know enough change management to actually teach it. Ironically, one of the Master Black Belt consultants was quoted as saying that Master Black Belts should not try to teach something they had never done in a real situation!

MK: I agree. For me, the addition of change management to my improvement efforts has helped give me more insights and options when something is not working and increases my ability to try something new rather than remain stuck in something. As Albert Einstein said, "The definition of insanity is doing the same thing over and over and expecting different results." The change management tools have helped me in my approach and to ask better questions that engage people, clarify issues, and better collaborate with my customer. It has also enabled me to create a positive environment for myself and others just by shifting my approach as translator from theory to application. I used to go in as the expert, but now I approach with questions and discovery of others' perspectives for a more informed understanding of the problem. That was huge.

RF: Using some change management methods in my work today helps me to better see beyond the initial technical problem and look at the issue holistically. The methods I have used have helped me to see all views of the problem and the options for action that are available beyond the technical solution. The tools have also given me a unique insight on how the area in the company I am working with got to where they are and why management behaves as they do. It is like I have a new set of lenses to see the world.

BF: "Okay, I will take the bait. Please tell me the change management tools you have been using and why those tools versus others?

RF: If you are talking about the stuff we learned around complexity and adaptive change tools that I have used, it would be Same/Different, CDE, Differences Matrix, Landscape Diagram, Simple Rules, and setting conditions for Self-Organization.

MK: I agree with Ron here, but I will add this twist—I use the tools where it appears to fit (it depends!) because not all tools are created equal in any given situation. The tools I have used are Simple Rules, Same/Different, Decision Map, Patterns, Self-organizing, and CDE.

RF: The impact of these tools and methods is that it gives you a way to be analytical and generate an unbiased perspective of what is needed or what needs to change without getting personal. The advantage lies in asking better questions. The methods I learned have helped me develop and maintain more objectivity with the sticky topics and issues that get in the way, and with conflict that arises, usually between two people or functions, that needs to be addressed in order to move forward. For example, a team recognized that it needed to take on a new role in coordinating the execution of projects spanning multiple development groups. The change analysis helped identify the place in the process to pilot their changes to best see if they would work (CDE, Landscape, and Self-organizing). The outcome was objective and the group moved forward.

MK: Let me think … what first comes to mind is the Decision Map as it helped discover the gap with one of the leaders I was working with and the team by bringing the unspoken differences to light and helped moved the project forward. It uncovered the poor assumptions about the project that were made because of past working relationships and freed up the team to see more objectively what needed to get done around alignment in order for the project to move forward. At minimum, the tools helped me better connect with my customers and offer a different viewpoint to approach problems. I will say honestly the results varied. Sometimes the customer didn't see the value or decided not to change. Other times, the tools offered a new perspective or missing piece of the puzzle that helped drive deeper discussions quickly that would have taken longer to understand. It all depended on where the customer was and to what degree they wanted to make a shift in really changing the work.

BF: Can either of you suggest the top methods or tools you think will help people reading this?

MK: That is hard since there is a lot and it all depends on the situation, but the ones I seem to use the most are Simple Rules, Same/Different, and Patterns. Many times when I am asked to support a team, it is because they want to do something different. Before one can decide if something different needs to be done, I help the team look at the patterns and determine if any patterns need to be changed. The Simple Rules and Same/Different help me to easily see the patterns and offer ways to help change those patterns.

RF: Let me explain what I know about these tools and methods from my experience. First, I use Same/Different when I see a change in a work process initiated (usually because one is needed). Often what needs to change is not well understood. The Same/Different discussion helps to focus the team and overcome any confusion and solve the problem faster. The Simple Rules are important and helpful because they align and give structure to the team and work situation so that it doesn't feel so chaotic. This is the power of simplification! It is well known that people often can't keep track of more than 7 to 10 things at once. Most of what happens in a company can be explained by a few (less than 10) simple rules (usually unwritten, too) of how to engage with each other. It is a social system just like the league of bowlers when they bowl. Simple rules give clearer guidance and enable better focus on the work and/or game.

MK: Whoa, Ron, very good on the explanation. Have you considered teaching? (Laughing) Please keep going as I am going to steal this stuff from you!

RF: Well, Okay then, I will. I like the Landscape Diagram because it can help explain and be a guide to self-organizing. It provides a simple model to help size up a group to help them more effectively see where they are stuck or need help or where they want to go. Knowing where the group is now is one piece of helping them decide where they need to go. Since I am on a roll, let me also add the CDE analysis, which helps the team get analytical about the team and work situation by giving names to containers, the differences between the containers and the exchanges that go on between the groups, and how they can be different. Naming the CDE stuff helps people see the issue together. Moreover, if the team does not name what is going on within the work, it gets ignored. Once it is named, the team and leaders have a chance to push a key difference and make change happen. This is critical in getting a group to self-organize. If leaders want to make change happen, provide the rules to the game so folks can play. If the rules don't work, believe me, the group will change them. You gotta love free will!

BF: Based on your experience now with Six Sigma and other process improvement efforts, if you were asked to coach senior leaders who wanted your help to roll out a companywide process improvement initiative across the organization, what advice could you offer? (And you can't say, "Don't do it.")

MK: Oh, Oh. I know this one. Before even starting, I would recommend that the leaders consider assessing first if the company has capacity for change. Are there too many things going on to add another initiative on top of overburdened employees? Have employees embraced change in the past or have they been able to wait out the leadership? Depending on the results, the initiative might need to be adjusted to be a major effort, secondary effort, or simply not do it at this time. The second data to consider is an assessment on past initiatives on what worked and what didn't, applying some of the tools for adaptive change we have already mentioned (same/different) and develop simple rules for what worked (or productive patterns). By assessing this capacity before doing a project, leaders can look at the culture or patterns of the organization and make informed decisions to continue and strengthen productive patterns and focus the change efforts to reduce negative patterns. In addition, I would recommend that anyone working on this improvement effort ensure that process improvement language and frameworks are translated into the experience or language of the organization. It will help spend less time discussing how to say things and get to the work sooner.

RF: I agree with the language points MK identified, but would like to add to move forward setting up Simple Rules on two levels. The first level is the public level. State the simple rules that relate to the business objectives. For example, focus on problems that are causing the most pain in the business, get help if you run into barriers, and help the group make the change their own. Next, leaders need to think ahead about what kind of other changes will be propagated by the initiative with questions like: "What other dominoes will fall when we knock this one over?" Then focus on the conditions necessary for the changes to be successful. What conditions need to be in place to help encourage change achieve the desired outcome? Failure to focus on the necessary conditions will only make the changes incremental and not transformative. For example, identifying the "talented" people as Black Belts created an unintended separation, and ill feelings emerged about Six Sigma experts being better than the people working the projects, which impacted the ability for creating real ownership and knowledge transfer beyond the Six Sigma organization of the Six Sigma skills and processes.

BF: Well, thank you both for your time and insight into how you apply adaptive change management to your work within Six Sigma and Lean.

21

What Next?

As we try to "close the circle" of this book, we'd like to recap the main points we introduced way back in Chapter 1:

- Far too many process improvement efforts fail to achieve their full potential, negatively impacting leaders, employees, customers, and shareholders. The problem is *not* with the process improvement tools themselves.
- The issue is embodied in the maxim: *"It ain't the what, it's the how."* While process improvement methods and tools are fundamentally sound in their own right, the way they have been implemented is not.
- We argued business process improvement implementation strategies have been focused on, and implemented with, the assumption that human systems are just like mechanical systems. This is a bad assumption.
- Human systems are *not* like mechanical systems. They do not respond the way machines do; they don't get fixed the way machines do; they're not interchangeable the way a machine's component parts are. Different models are needed to address these human dynamics.

In Section II, a case story was introduced to dramatize the dynamics of a linear Six Sigma implementation. That section closed by asking you, the reader, to notice what was similar or different in your own experience as you learned about *TryinHard Marine*. Is (was) your organization guilty of:

1. Looking to Lean Six Sigma as a "quick fix" for all of their problems?
2. Not taking their prior history with process improvement into account?
3. Abdicating leadership responsibility and accountability?
4. Using incomplete selection criteria in their selection of a consultant?
5. Poorly identifying and scoping their initial projects?
6. Using the wrong criteria in selecting people for the initial projects?
7. Not investing in leadership training?
8. Putting too much emphasis on Six Sigma tools training?
9. Leaders launching the initiative and not following through?
10. Failing to engage the middle management of the organization?
11. Failing to understand or address conditions of the market, organization, or process for change to happen?
12. Not having strategies to adapt to the inevitable changes that occurred in the organization during implementation?

We then retold the case story in Section III, this time integrating adaptive change tools that recognize the nonlinear, complex properties of human organizations. In the flow of the story, you learned:

1. Linear, "cause and effect" implementation models need to be supplemented with the concept of nonlinear *patterns*. Organizations have to let go of their paradigm of guaranteed results of change efforts.
2. The *Eoyang CDE model*, where we demonstrated how making a change in **C**ontainers (formal and informal groupings), **D**ifferences (amplifying or damping differences or tensions), or **E**xchanges (the vehicles for exchanging tangible goods, information, and energy) can influence systemic patterns. We showed how changing any one element of the Eoyang CDE model impacted the others, though not necessarily in predictable or expected ways.
3. The *System Landscape* where we introduced the concept that different elements of complex systems can reside in a space best characterized as *Organized* (high degree of certainty and agreement), *Unorganized* (low degree of certainty and agreement), or *Self-Organizing* (an intermediate degree of certainty and agreement that provides some amount of structure while allowing for the flexibility needed to adapt to changes in the system or its surrounding environment).

4. The *7 Cs Model*, which defines seven areas to consider when evaluating the "dynamic fit" of a system (or project) with its current and projected future surroundings. Each of these 7 Cs: **C**onnections (linkages with other parts of the system), **C**ommunication (transfer of information, communication channels), **C**ontinuous Assessment (measurement, feedback, "How do we assess how we're doing?"), **C**ommitment (ownership, buy-in), **C**onstructs (physical manifestations, evidence), **C**ongruence (alignment), or **C**apacity (skills, competencies, and resource levels) provides a shared language for conversation and a potential opportunity for an intervention.

5. *Adaptive action:* a rapid, iterative process that is consistent with a complex adaptive view of the world. Beginning with "What?" (describing the current state), then asking "So what?" (determining what is most important of the data collected; identifying the most promising levers for change), adaptive action concludes by asking "Now what?" (identifying what interventions might influence the pattern in desired ways). Noticing the results of the "Now what?" intervention starts the adaptive action cycle all over again.

6. *Simple Rules* are the critical few agreements about behavior that allow individual members of the organization to act with the required balance of structure and flexibility to be adaptive. So long as everyone in the system knows the simple rules and follows them, the patterns of the system can adapt to change in productive ways.

7. *Decision Maps* that recognize conflicts between members of a complex system are often due to an issue in one of three areas: (1) different *world views* (paradigm, philosophies, values, etc.); (2) the *reality* parties in a conflict pay attention to (data); or (3) the *rules* or strategies each follows to resolve differences in specific situations. By surfacing these (usually subconscious) beliefs, parties in a disagreement can better understand the nature of the conflict. By creating an agreement about at least one of the three areas, the pattern can be shifted, positively influencing the conflict situation.

8. *Comparing reality and perceptions* is another tool for allowing new possibilities for action to appear in a conflict situation. By separating the *data each member of a conflict pays attention to* from the *perception each party has about that data*, differences in patterns can be identified and different alternatives for resolution identified.

9. The *Fractal* properties of many complex adaptive systems provide another tool for adaptation and self-organization. By recognizing that the essential properties of fractal systems are scale-free, that is, they retain certain characteristics at both macro and micro levels, new opportunities for influencing behavior change appear. These fractal properties help explain how Skunk Works® or other changes created at the edge of organizations can sometimes be amplified to create large changes throughout the entire system.

Having acquainted you with these new "tools for your toolbox," we shift to a different, higher level of abstraction (from 5,000 ft. to 25,000 ft.) to make some wide angle lens observations about implementing process improvement approaches in complex adaptive systems.

1. **Initial conditions matter:** In keeping with our mantra throughout this book that "one size does not fit all," from where, specifically, are you and your organization starting? Was your prior experience with change good or bad? Where are your customers right now with regard to the proposed change? What is the current skill level of your employees? How much energy is available to take on the work of this improvement process?
2. **The Golden Rule does not always apply:** The old adage, "treat other people as you would like to be treated," is not always helpful. It assumes that others think, feel, and act in the world the same as you, have the same world view, focus on the same data, and follow the same rules. In complex systems, not only is that not likely to be true, it could be harmful if it dampens the differences that allow flexibility, adaptation, and self-organization to occur. We advocate the Platinum Rule instead, *"treat others as they would prefer to be treated."* This will sensitize people to the differences around them as well as the similarities.
3. **Change is essential for survival:** The process of change often gets a bum rap. It can be hard, difficult, time and resource-consuming. Through the lens of complex adaptive systems, though, change is essential for survival and growth. What happens to a complex adaptive system, say, a human being, if her heart does *not* change, if it doesn't expand and contract? What happens if a person's bloodstream

maintains the status quo, doesn't change, doesn't capture oxygen from the lungs and distribute it to the rest of the body?

4. **Tread carefully as a change agent:** We have learned this over and over again in our work with clients at every management level, in companies large and small. We complex human beings have these inconvenient, messy things called emotions and egos. We don't like to look bad or foolish while learning something new. We don't like to lose in the (often self-defined) competition with others to be the "first and best" at what is required by a new change. We don't always respond well to the fear that we might not be successful in the new way.

   Everyone engaged in the work of process improvement, whether leader, sponsor, team leader, or team member is a change agent, embedded in and creating the "messy" situations we just described. Remember, people are not machines. They bring these feelings with them, despite the signs posted at the door to check your feelings at the door. Remaining sensitive to these very real aspects of a change process will hopefully influence both your approach and tools.

5. **There are no guarantees:** And yes, that most certainly includes the perspectives, concepts, and tools we've introduced in this book. For us to promise that these tools will work would be to fall into the same trap we've railed against throughout this entire story. There are some things we are reasonably confident about. One is: *"If you do what you've always done, the way you've always done it, it's naïve to believe you'll get a very different result."* Another is: *"Systems are perfectly designed to get the results they're getting right now."* Look to your past and current experience with business process improvement. If you didn't achieve the results you hoped for, try something different.

In closing, as much as we would like to, we can't take the next step for you—the "Now What?" Folk wisdom says, *"You can lead a horse to water, but you can't make it drink."* We think that applies here. Please, **try something (anything)** of what we've shared in this book. Think about it as an experiment. The "null hypothesis" (remember that from high school scientific method class?) is "no significant difference." What is the worst that can happen? Your experiment won't work the way you expect. The way we see it, so long as you notice the result and learn from it, it was worth doing.

We hope that *What Works for GE May Not Work for You: Using Human Systems Dynamics to Build a Culture of Process Improvement* provides new possibilities for making change happen. The authors continue to explore these concepts and hope to remain in learner mode for a long time to come. We invite you to share your questions and experiences with us and with other readers and interested bystanders. Please visit our website: *www.whatworksforgemaynotworkforyou.com*. Add your own experience or read about the experiences of others. Pooling our efforts, we can create a self-organizing system that can help each of us improve our dynamic fit with the changing world that surrounds us.

Happy experimenting!

Recommended Reading

In the area of Human Systems Dynamics, Chaos, Complexity, and Systems:

Briggs, John and Peat, F. David, *Turbulent Mirror: An Illustrated Guide to Chaos Theory and the Science of Wholeness.* 1990, Harper & Row: New York.

Davis, Stan and Meyer, Christopher, *Blur: The Speed of Change in the Connected Economy.* 1998, Addison-Wesley: Reading, MA.

Denning, Stephen, *The Springboard: How Storytelling Ignites Action in Knowledge-Era Organizations.* 2001, Butterworth-Heineman: Woburn, MA.

Eoyang, Glenda H., *Coping with Chaos: Seven Simple Tools.* 1997, Lagumo: Cheyenne, WY.

Eoyang, Glenda, Ed., *Voices from the Field: An Introduction to Human Systems Dynamics.* 2003, Human Systems Dynamics Institute Press: Circle Pines, MN.

Gleick, James, *Chaos: Making a New Science.* 1987, Penguin Books: New York.

Goldstein, Jeffrey, *The Unshackled Organization: Facing the Challenge of Unpredictability through Spontaneous Reorganization.* 1994, Productivity Press: Portland, OR.

Holladay, Royce and Quade, Kristine, *Influencing Patterns for Change: A Human Systems Dynamics Primer for Leaders.* 2008, Human Systems Dynamics Institute Press: Circle Pines, MN.

Lewin, Roger, *Complexity: Life at the Edge of Chaos.* 1992, Macmillan Publishing Co.: New York.

Marrow, A.J., *The Practical Theorist: The Life and Work of Kurt Lewin.* 1977, Teachers College Press: New York.

Olson, Edwin E. and Eoyang, Glenda H., *Facilitating Organizational Change: Lessons from Complexity Science.* 2000, Jossey-Bass/Pfeiffer: San Francisco.

Oshry, Barry, *Seeing Systems: Unlocking the Mysteries of Organizational Life.* 1996, Berrett-Koehler: San Francisco.

Sanders, T. Irene, *Strategic Thinking and the New Science: Planning in the Midst of Chaos, Complexity, and Change.* 1998, The Free Press: New York.

Senge, Peter M., *The Fifth Discipline: The Art and Practice of the Learning Organization.* 1990, Doubleday Dell Publishing: New York.

Senge, Peter M., Kleiner, Art, Roberts, Charlotte, Ross, Richard B., and Smith, Bryan J., *The Fifth Discipline Fieldbook: Strategies and Tools for Building a Learning Organization.* 1994, Doubleday Dell Publishing: New York.

Stacey, Ralph D., *Managing the Unknowable: Strategic Boundaries between Order and Chaos in Organizations.* 1992, Jossey-Bass: San Francisco.

Waldrop, M. Mitchell, *Complexity: The Emerging Science at the Edge of Order and Chaos.* 1992, Simon and Schuster: New York.

In the area of Lean Enterprise and Six Sigma:

Brassard, Michael and Ritter, Diane, in partnership with Sterett, Kent, *Sailing through Six Sigma.* 2001, Brassard and Ritter, LLC: Marietta, GA.

Breyfogle III, Forrest W., *Implementing Six Sigma: Smarter Solutions Using Statistical Methods.* 2003, John Wiley & Sons: Hoboken, NJ.

Eckes, George, *The Six Sigma Revolution: How General Electric and Others Turned Process Into Profits.* 2001, John Wiley & Sons: Hoboken, NJ.

Flinchbaugh, Jamie and Carlino, Andy, *The Hitchhiker's Guide to Lean: Lessons from the Road.* 2006, Society of Manufacturing Engineers: Dearborn, MI.

George, Michael L., *Lean Six Sigma: Combining Six Sigma Quality with Lean Speed.* 2002, McGraw-Hill: New York.

Goldratt, Eliyahue M. and Cox, Jeff, *The Goal: A Process of Ongoing Improvement.* 1986, North River Press: Croton-on-Hudson, NY.

Greif, Michel, *The Visual Factory: Building Participation through Shared Information.* 1991, Productivity Press: Portland, OR.

Keyte, Beau and Locher, Drew, *The Complete Lean Enterprise: Value Stream Mapping for Administrative and Office Processes.* 2004, Productivity Press: New York.

Macinnes, Richard L., *The Lean Enterprise Memory Jogger: Create Value and Eliminate Waste throughout Your Company.* 2002 GOAL/QPC: Salem, NH.

Mann, David, *Creating a Lean Culture: Tools to Sustain Lean Conversions.* 2005, Productivity Press: New York.

Pande, Peter and Holpp, Larry, *What Is Six Sigma?* 2002, McGraw-Hill: New York.

Pande, Peter S., Neuman, Robert P., and Cavanagh, Roland R., *The Six Sigma Way: How GE, Motorola, and Other Top Companies Are Honing Their Performance.* 2000, McGraw-Hill: New York.

Productivity Press Development Team, *The Lean Office: Collected Practices and Cases.* 2005, Productivity Press: New York.

Pyzdek, Thomas, *The Six Sigma Handbook: A Complete Guide for Greenbelts, Blackbelts, and Managers at All Levels.* 2001, McGraw-Hill: New York.

Schonberger, Richard J., *Japanese Manufacturing Techniques: Nine Hidden Lessons in Simplicity.* 1982, The Free Press: New York.

Shingo, Shigeo, *Non-Stock Production: The Shingo System for Continuous Improvement.* 1988, Productivity Press: Cambridge, MA .

Womack, James P. and Jones, Daniel T., *Lean Thinking: Banish Waste and Create Wealth in Your Corporation.* 1996, Simon & Schuster: New York.

Index